D1001414

ELISHA BARTLETT'S PHILOSOPHY OF MEDICINE

Philosophy and Medicine

VOLUME 83

Editors

H. Tristram Engelhardt, Jr., *Center for Medical Ethics and Health Policy, Baylor College of Medicine and Philosophy Department, Rice University, Houston, Texas.*

S. F. Spicker, *Massachusetts College of Pharmacy and Allied Health Sciences, Boston, Massachusetts.*

Associate Editor

Kevin Wm. Wildes, S.J., *Department of Philosophy and Kennedy Institute of Ethics, Georgetown University, Washington, D.C.*

CLASSICS OF MEDICAL ETHICS 2

Series Editors

Laurence B. McCullough, *Center for Medical Ethics and Health Policy, Baylor College of Medicine, Houston, Texas*

Robert B. Baker, *Department of Philosophy, Union College, Schenectady, New York and Center for Bioethics, University of Pennsylvania, Philadelphia, Pennsylvania*

H. Tristram Engelhardt, Jr., *Center for Medical Ethics and Health Policy, Baylor College of Medicine and Philosophy Department, Rice University, Houston, Texas*

S.F. Spicker, *Massachusetts College of Pharmacy and Allied Health Sciences, Boston, Massachusetts*

The titles published in this series are listed at the end of this volume.

Elisha Bartlett's Philosophy of Medicine

Edited by

William E. Stempsey, S.J.
College of the Holy Cross,
Worcester, MA, U.S.A.

MIDDLEBURY COLLEGE LIBRARY

A C.I.P. Catalogue record for this book is available from the Library of Congress.

ISBN 10 1-4020-3041-X (HB)
ISBN 10 1-4020-3042-8 (e-book)
ISBN 13 1-4020-3041-3 (HB)
ISBN 13 1-4020-3042-0 (e-book)

Published by Springer,
P.O. Box 17, 3300 AA Dordrecht, The Netherlands.

www.springeronline.com

Printed on acid-free paper

All Rights Reserved
© 2005 Springer
No part of this work may be reproduced, stored in a retrieval system, or transmitted in any form or by any means, electronic, mechanical, photocopying, microfilming, recording or otherwise, without written permission from the Publisher, with the exception of any material supplied specifically for the purpose of being entered and executed on a computer system, for exclusive use by the purchaser of the work.

Printed in the Netherlands.

TABLE OF CONTENTS

PREFACE ix

PART I

William E. Stempsey, S. J. / INTRODUCTION 1

CONTENTS OF THIS BOOK 2

A BIOGRAPHICAL SKETCH OF ELISHA BARTLETT 3

Early Life and Education 3

Medical Education 3

Life in Lowell 5

The Peripatetic Professor 6

Return to Rhode Island 10

PHILOSOPHY OF MEDICINE IN THE NINETEENTH CENTURY 11

Early Nineteenth Century American Medicine 12

The Paris Clinical School 14

Influence of the Paris Clinical School in America 15

ELISHA BARTLETT'S PHILOSOPHY OF MEDICINE 16

An Introductory Lecture on the Objects and Nature of Medical Science 17

An Essay on the Philosophy of Medical Science 18

An Inquiry into the Degree of Certainty of Medicine, and into the Nature and Extent of Its Power over Disease 26

The Philosophy of Therapeutics 29

BARTLETT'S OTHER WRITINGS 31

Editorial Work and Translation 32

Medical Writing 32

Medical Exhortation 33

Civic Addresses 36

Poetry 37

BIBLIOGRAPHY 41

PART II

Elisha Bartlett / TEXTS

An Essay on the Philosophy of Medical Science 49
Lea and Blanchard, Philadelphia, 1844.

Part First: Philosophy of Physical Science 58

Chapter I 60

Chapter II 62

Chapter III 69

Chapter IV 71

Chapter V 80

Part Second: Philosophy of Medical Science 87

Chapter I 89

Chapter II 91

Chapter III 95

Chapter IV 96

Chapter V 97

Chapter VI 100

Chapter VII 106

Chapter VIII 107

Chapter IX 115

Chapter X 118

Chapter XI 126

Chapter XII 140

Chapter XIII 160

Chapter XIV 171

Chapter XV 181

Chapter XVI 185

Index 199

The Philosophy of Therapeutics 201
Manuscript, ca. 1852

William E. Stempsey, S. J. / NOTES 213

INDEX 235

PREFACE

The idea of preparing a new critical edition of Elisha Bartlett's *Essay on the Philosophy of Medical Science* was suggested to me several years ago by Dr. H. Tristram Engelhardt, Jr. Since that time it has been a pleasure to get to know the life and work of Elisha Bartlett. I am pleased to be completing this book in the bicentennial year of Bartlett's birth.

Bartlett was born in 1804 in Smithfield, Rhode Island, less than twenty-five miles from Worcester, Massachusetts, my present home—a short journey even in Bartlett's day. I have been able to walk at some of the sites to which Bartlett continually returned during his life. Visiting Bartlett's grave in the Slatersville cemetery has been an inspiration for the preparation of this book.

Proximity to several institutions with rich holdings in Bartlett's works and in nineteenth-century American history of medicine greatly facilitated my research. First, though, I want to acknowledge the College of the Holy Cross for supporting my sabbatical leave for the academic year 2003-2004. The American Antiquarian Society, in Worcester, Massachusetts, was generous in giving me access to its remarkable resources. I was able to find many of Bartlett's published works and other nineteenth-century medical literature there, and the entire library staff provided quick and able research assistance. For access to other published works, I am indebted to the Medical Historical Library of the Harvey Cushing/John Hay Whitney Medical Library at Yale University and the Department of Rare Books and Special Collections of the Countway Library of Medicine at Harvard Medical School. The staff at both of these institutions also provided valuable assistance. The Thomas P. O'Neill Library at Boston College and the Lamar Soutter Library at the University of Massachusetts Medical School were excellent and convenient sources of nineteenth-century medical literature on microfilm.

Several libraries also generously granted me permission to examine the papers and manuscripts of Elisha Bartlett. Two major collections of Bartlett's papers are held by the Manuscripts and Archives division of the Sterling Memorial Library at Yale University and the Department of Rare Books and Special Collections of the Rush Rhees Library at the University of Rochester. A few of Bartlett's letters are also contained in the Oliver Wendell Holmes papers in the Houghton Library at Harvard University, and the Oliver Wendell Holmes papers and Edward Everett papers in the John Hay Library at Brown University. Bartlett's *Philosophy of Therapeutics*, published in this volume, was pieced together from undated manuscript fragments in the collections of Yale and Rochester. Permission to publish this material and to quote from correspondence was kindly granted by the following: Brown University Library; the Houghton Library, Harvard University;

the Department of Rare Books and Special Collections, University of Rochester Library; and Manuscripts and Archives, Yale University Library.

Elisha Bartlett and the people to whom he wrote seemed to be aware that future generations might be interested in his writings; they preserved an astounding number of his letters and manuscripts. I make no attempt in this book to catalog all of Bartlett's correspondence, since it touches upon the many facets of Bartlett's relatively short but extraordinarily varied life—not only medicine, but also poetry, politics, and travel. The Yale collection contains Bartlett's lecture notes from medical school days and an extensive set of letters written to his sister during his studies in Europe, mainly Paris, in the year following his graduation from medical school. The Rochester collection contains many family documents and personal artifacts, including Bartlett's medical diploma given by Brown University in 1826. This book highlights the manuscript material that is most directly related to Bartlett's philosophy of medicine.

Finally, I thank my home community, the Jesuit Community of the College of the Holy Cross, for support of my travel. The distances were not very long, and an automobile made my travel much easier than the travel of Elisha Bartlett, who went much greater distances by stagecoach and riverboat. Nonetheless, I would not have been able to work in such lovely settings with rare books and manuscripts without this support. I also thank the Jesuit communities at Fairfield University in Fairfield, Connecticut, and at McQuaid Jesuit High School in Rochester, New York for their generous hospitality. All these fellow Jesuits provided me not only with means to travel and a place to stay, but also with fine companionship at home and on the road.

Worcester, Massachusetts
July 2004

INTRODUCTION

Elisha Bartlett (1804-1855) was a physician, politician, poet, philosopher, and a prominent "peripatetic professor." In a relatively short life, this Rhode Island native established a medical practice, taught in eight medical schools from New Hampshire to Kentucky, and served as the first mayor of Lowell, Massachusetts and as a Massachusetts legislator. He was a noted public speaker; many of his addresses, some on the controversial issues of his day and some merely celebrating some festive occasion, were published as pamphlets for the public. Bartlett defended the health of the young women working in the mills of Lowell. He was an abolitionist. His poetry was printed in several periodicals and he published a small book of poetry later in his life. He was a friend of Oliver Wendell Holmes and an acquaintance of Charles Dickens, with whom he corresponded and exchanged work.

It is Bartlett's philosophy of medicine, however, that is the major topic of this book. Bartlett's epitaph comes from Job 29:11, Job's final soliloquy, in which he speaks of past happiness. Although it was well known and often used in the nineteenth century, it is an apt expression of Bartlett's empiricist epistemology: "When the ear heard him, then it blessed him, and when the eye saw him, it gave witness to him." Bartlett's philosophy of medicine was iconoclastic in his medical world. It is not surprising that it should be controversial, then and now. William Osler, speaking of Elisha Bartlett, says that Rhode Island "can boast of one great philosopher, one to whose flights in the empyrean neither Roger Williams nor any of her sons could soar" (15). Lester King is less kind, calling Bartlett "arrogant and narrow-minded" ("Medical Philosophy," 156), and Bartlett's philosophy "third rate" (*Transformations*, 206). It is true that Bartlett's empiricism does not approach the sophistication of what had already been articulated by philosophers such as John Locke and David Hume and that Bartlett was apparently not acquainted with the epistemological revolution occasioned by Immanuel Kant. It is also true that nothing of Bartlett's philosophy of medicine is particularly original. Nonetheless, such severe criticism may be unfair. King ("Medical Philosophy," 156) even admits that Bartlett's *Essay on the Philosophy of Medical Science* is "a work of considerable importance for understanding the medical thought of the nineteenth century." Indeed, it is probably the best presentation of the philosophy of the Paris clinical school of medicine that became so prominent in nineteenth-century America. For this reason alone, Bartlett's work deserves continued attention.

The early nineteenth century was an intriguing time for medicine. It was, in many ways, a golden era for the philosophy of medicine. It was a time of fierce competition between so-called sects in medicine. The "regular" physicians competed for patients with the Thomsonians, Eclectics, and homeopaths (Starr, 93-

99). Even in the ranks of the regular physicians, disputes over pathology and therapy reflected thought-provoking differences in philosophy of medicine. Elisha Bartlett was not afraid to grapple with the issues of medical philosophy that so divided nineteenth-century medicine. His entry into the fray, not surprisingly, put him at odds with some of the most influential medical practitioners and thinkers of his day.

CONTENTS OF THIS BOOK

This book is primarily conceived as a critical edition of Bartlett's 1844 *Essay on the Philosophy of Medical Science*. As the reader will discover, Bartlett's *Essay* was seen in its day as an important contribution to medical philosophy by a well known and respected physician and educator. The present volume makes Bartlett's work available to scholars. The text, notes and front matter are all provided. Page numbers of the original edition are given in brackets at the end of the text on the original page. Original spelling has generally been retained; only obvious printing errors are corrected, and these corrections are noted. Likewise, the often unconventional original punctuation has been retained. One change has been made in Bartlett's footnotes. In the original edition, notes are numbered beginning anew on each page. The present volume retains Bartlett's notes as footnotes, but marks them with lower-case letters, beginning anew with each of Bartlett's chapters. Editorial notes are given as endnotes, marked with Arabic numerals.

Also included in this book is Bartlett's previously unpublished *Philosophy of Therapeutics*. This short essay was gleaned from undated manuscript fragments, all written in Bartlett's own hand with the exception of one small addition to the text. It apparently is a development of a lecture that Bartlett gave around 1852; the introduction to the lecture, not included in this edition of the essay, refers to Bartlett's chair of materia medica and medical jurisprudence, which is the position he took at the College of Physicians and Surgeons in New York in 1852. Bartlett lectured there for only two sessions before he became ill and was forced to retire to his home in Rhode Island. Therefore, the manuscript on the philosophy of therapeutics must have been written between 1852 and 1855, when Bartlett died. This short essay reiterates Bartlett's empiricist project, but it also further develops a few of the points made in the *Essay on the Philosophy of Medical Science*. It provides us with a more complete picture of Bartlett's philosophy of medicine, adding a philosophy of an important matter of medical practice to his already published philosophy of medical science.

This introduction sets the context for Bartlett's philosophical writings and offers some critical reflection on Bartlett's positions. Following a biographical sketch of Elisha Bartlett, there is a discussion of the philosophy of medicine in the nineteenth century and Bartlett's contribution to it. A bibliography of Bartlett's works and secondary literature on Bartlett and his philosophy of medicine is included. It is hoped that this new edition of Bartlett's philosophy of medicine and the bibliography will be valuable to scholars of the philosophy and history of medicine.

A BIOGRAPHICAL SKETCH OF ELISHA BARTLETT

Although there is no book-length biography of Elisha Bartlett, biographical sketches are given by Holmes, Dickson, Rider, Osler, The National Cyclopædia of American Biography, Burrage, and Miller. It is from these accounts and from Bartlett's papers and manuscripts that I draw this sketch.

Early Life and Education

Elisha Bartlett was born in Smithfield, Rhode Island on October 6, 1804, son of Otis and Waite Buffum Bartlett, and descended from John Bartlett, who was born in Weymouth, Massachusetts in 1666. Elisha Bartlett's grandfather, also named Elisha, came to Branch Village, a section of Smithfield, from Glocester, Rhode Island in 1795 and began the manufacture of scythes. Otis Bartlett carried on the scythe-making business (T. Bartlett, 95). Otis and Waite Bartlett were members of the Society of Friends, and raised Elisha in that tradition. Bartlett apparently did not continue to worship as a Quaker in adulthood. He was one of the prominent organizers of the Unitarian Church in Lowell in 1829 (Coburn, vol. 1, 199). He also talks of his interest in Unitarianism in a letter to George Bartlett from England in 1846 (Y1:10:5). Even so, the Quaker influence was, in the view of Huntington (25-26), an influence throughout Bartlett's life in the "quietness of his demeanor, in the moderation of his views, and the gentleness of his spirit."

Details of Bartlett's early education are sketchy, but in a letter to Elisha Huntington, M.D., Bartlett's brother George reports that Elisha Bartlett attended a seminary in his native town, a school in Uxbridge, Massachusetts, and the well-known Friends' institution in Nine Partners, New York, where he studied under the care of Jacob Willets (R1:18:1). During Bartlett's boyhood, books "were his companions, his solace and his delight," and it is said that he received as fine a classical education as any American university at the time could give (Huntington, 5). The erudition and grace of many of Bartlett's writings, even some of his student days (R2:1:1), attest to this.

Medical Education

Bartlett's medical education might also be characterized as being very fine for the time, although unsystematic by current standards. In the early nineteenth century, becoming a physician who evinced the highest standards required thirty-six months of consecutive study. This included a two-year apprenticeship, and two sessions of medical school lectures, each session lasting thirteen to sixteen weeks. The second session was typically just a repetition of the same set of lectures. Students could attend more than one medical school, however, and so fulfill the two-session requirement within one calendar year. Finally, the student would take the examinations and complete a thesis, which was often rudimentary (Bonner, 175-81;

Starr, 42-44; Waite, 20-21, 27). Bartlett pursued studies with several distinguished physicians: Dr. George Willard of Uxbridge, Massachusetts; Dr. John Green and Dr. B. F. Heywood of Worcester, Massachusetts; and Dr. Levi Wheaton of Providence, Rhode Island. He also took up other duties during his apprenticeship years. On November 16, 1824, he wrote from Worcester to his sister: "I have taken up my place of abode with Judge Lincoln–and shall probably remain in his family as long as I have the care of his boys" (R1:2:1).

Bartlett attended medical lectures in Boston and Providence. His lecture notes from the courses given by Dr. James Jackson (Y2:16:1) and Dr. Jacob Bigelow (Y2:17:1) in Boston during the 1824-25 term are extensive and provide insight into the state of medicine at the time. A letter from Bartlett, dated November 20, 1825, from Boston to his sister Caroline (R1:2:2) tells a bit about the life of a medical student at the time.

> My boarding house is at Mrs. McCowdray's, no. 3½ Sudbury Street. . . . There are two medical students here, beside myself, and we all occupy one room.—Our landlady keeps an excellent table, and charges but $3 per week—We usually hear five lectures each day of an hour each. We visit the Hospitals twice a week, once at ½ past 8 in the morning, and once at 12.

Bartlett received the degree of Doctor of Medicine from Brown University in 1826, a year before that university closed its medical department for nearly a century and a half.

In June 1826, Bartlett journeyed to Europe, near the beginning of the wave of American medical students who furthered their medical studies in Paris. His letters to his sisters (Y1:1:1, Y1:1:2, Y1:1:3, Y1:2:2) give an extensive account of his travels, but say relatively little about his medical studies. Bartlett stayed in Paris, where he heard medical lectures of Cloquet, Cuvier, St. Hillaire, Dupuytren and others, and attended the practice of the hospitals. This stay would be of utmost importance for the development of Bartlett's philosophy of medicine. Indeed, for Bartlett and the many American physicians who followed him to study in Paris, the French practice of medicine would become the model for the development of American medicine. In a letter of September 27, 1826, to his sisters (Y1:1:2), he reports that he has commenced a two and one-half month course of operative surgery. He writes: "We perform operations ourselves, in addition to seeing them performed by the lecturer." Bartlett also comments on the familiarity of the French surgeons and physicians with their students—a stark contrast to Bartlett's experience in New England. Bartlett attended public lectures beginning early in the morning and lasting late into the evening, visiting the hospitals by candlelight and attending lectures there immediately after the visits. Oddly, there is no mention of any contact with Pierre Louis, whose philosophy of medicine so greatly influenced Bartlett and to whom Bartlett dedicated his *Essay on the Philosophy of Medical Science*. It is uncertain whether Bartlett studied with Louis at this time or whether he simply became acquainted with the work of Louis through lectures and reading. Bartlett remained in Paris until December 1826, and then journeyed to Italy, where he visited the classical sites and marveled in the works of art. He returned to Paris in

March 1827, spent the month of May in London, and sailed for home from Liverpool on June 8, 1827.

Although the letters are often written in haste so that they could be delivered to sailing ships, they are quite literate. The meticulous character of Elisha Bartlett comes through in his writing. He was obviously one who took pride in how he presented himself. This is evident from a letter written from Leghorn, Italy on December 28, 1826 (Y1:1:3), in which he talks about the poetry of Lord Byron:

> You and those of my friends who take the trouble to read my letters—must recollect—that they are necessarily written with haste which forbids all arrangement—or care—that they are seldom even read over—therefore be indulgent—The matter may render some of them of some interest—but I intreat [*sic*] you to let them never go unguarded into the hands of an Editor—What I have said of Byron I should regret exceedingly to see in a publick print.

Regarding the comments in this letter about Byron, this editor will respect Bartlett's wishes.

Life in Lowell

Upon returning to America in 1827, Bartlett took up residence in Lowell, Massachusetts, a town that was rapidly growing because of the establishment there of numerous mills. He began a medical practice and developed a great concern for the physical well-being of the town's citizens. This extended beyond Bartlett's work as a physician and led him to take an active role in the civic life of Lowell. In 1828, when he was only twenty-four, he gave a lecture on contagious diseases before the Lowell Lyceum; in that same year he was the Fourth of July orator. He delivered the address at the dedication of Mechanics Hall in 1835. In 1836, he gave a course of lectures on physiology, which drew a large public turnout. Also in 1836, he received an invitation given only to a few, to deliver an address at the Odeon in Boston (Patterson, 366).

In 1829, Bartlett married Elizabeth Slater, of Smithfield, Rhode Island. Elizabeth was the daughter of John Slater (for whom Slatersville, their section of North Smithfield, was named) and the niece of Samuel Slater, who brought the Industrial Revolution to America with his mill in Pawtucket, Rhode Island. There is no record of their having any children.

When the town of Lowell was incorporated as a city in 1836, Bartlett, running as a Whig, was elected its first mayor, defeating the Reverend Eliphalet Case, 958-868. Bartlett was elected for a second term in 1837, but after completing that term he "declined all further service in this line." (Cowley, 167). He did, however, represent the Lowell district in the Massachusetts legislature in 1841 (Cowley, 224). Huntington (8) remarks that this career was never marked by any extraordinary success. Rider (7) says that "these positions were uncongenial to his delicate nature, and he soon abandoned them for a more congenial pursuit, that of a medical teacher." The latter comment is set in a work of almost hagiographic tone, however. Bartlett's letters suggest that he was straightforward and not at all retiring or prissy.

Certainly, the travels he would undertake, going to the developing American West, would not have been congenial to someone with a delicate nature.

The Peripatetic Professor

The list of Bartlett's academic appointments is a long and complex one. Some preliminary comments on this are in order. As we have already noted, medical school sessions typically lasted thirteen to sixteen weeks. "Country" medical colleges began to proliferate in early nineteenth-century America, especially as the nation expanded westward. Many of these schools were in small towns in rural areas (Bonner, 179). The faculty typically consisted of only six or seven professors, only a couple of whom were resident faculty. The rest simply visited for the sessions, gave their lectures, and moved on. While a few of these teachers were wealthy, most were not; they saw teaching as a means to additional income and social prestige (Bonner, 89). In the history of the medical college at Woodstock, Vermont, twenty-four of its thirty-one faculty members were visiting professors (Waite, 24). The session of most of the urban medical schools ran from November until February, without a break for Christmas. These were known as the "winter medical colleges." Others, such as Dartmouth, were "fall medical colleges," with sessions running from August until November. Still others were "summer medical colleges," with sessions from June until September. Given these staggered schedules, the "peripatetic professors" often held positions in two medical schools. Some even taught in three or four different schools during the course of a calendar year (Waite, 25-26).

Elisha Bartlett was a prominent peripatetic professor. His first teaching appointment came in 1832, as professor of pathological anatomy and materia medica at the Berkshire Medical Institute in Pittsfield, Massachusetts. At the time this medical school was considered to be a very strong one, with classes of over one hundred (R1:4:2). This would have rivaled the schools in Boston. Bartlett held this chair for eight sessions, making his tenure in Pittsfield partially overlap his term as mayor of Lowell. Bartlett seemed to be quite happy in Pittsfield. His letters to Dr. John Orne Green, of Lowell, tell us a bit about the school and Bartlett's part in it. On November 25, 1833 (R1:3:1), Bartlett wrote:

> The character of the class is said to [be] superior even to that of last year. We have a large number of excellent students. . . . I have lectured, most of the time, twice a day,—at 10 a.m. and at 2 p.m. I shall finish my course on materia medica by the middle of this week, and the remainder of my time will be occupied with lectures of medical jurisprudence and pathological anatomy. . . .
>
> Our students have famous times here every Thursday in the discussion of medical questions. One member of the society reads a dissertation, and then follows the discussion. The question on the first evening at which I was present was this—"Is the primary function of the liver that of depurating the blood?" The question on the second eveg. was— whether there be or be not such a disease as an idiopathic fever without any concomitant local affection—The question last Thursday eveg. was this—Is the Negro race susceptible of the same mental cultivation as the Caucasian?

Two years later, Bartlett wrote again to Green (R1:4:1). He talks about a good class, and says that he is giving pathological lectures. (Presumably, he means lectures about pathology.) One of the benefits, it seems, of the role of peripatetic professor is the luxury of being away from the routine course of one's normal business for a period of several months. This leisure seems to have been a particular delight to Bartlett.

> I pass a good proportion of my time, here, quite alone, so that I find myself whiling away the hours in meditation much oftener than when I am engaged in the more varied and active affairs of business, at home. I think that I always leave Pittsfield with the better and purer part of my being somewhat strengthened.

Wanderlust seems to have begun to get hold of Bartlett, however. In 1838, he wrote from Pittsfield to his "dear friend" about an appointment to a professorship at Dartmouth, and also about "intimations" of a place on the medical faculty of the "new University of the City of New York." Bartlett says: "Where, or under what auspices I shall settle down in life, seems now to be very uncertain" (Y1:4:1). In 1839, Bartlett was appointed to the chair of practice at Dartmouth College. On the eighth of September in that year, he wrote to his friend Green from Hanover, New Hampshire (R1:5:1):

> All is getting on here, on the whole, pretty well. The class is respectable, and that is about all that can be said. . . . I have given 44 lectures:—have got thro' with the chest and head; from which I go tomorrow to the abdomen—to wind up with fevers. I chose to follow this arrangement, rather than the usual plans, because it enables me to treat first of the simpler, less complicated, more intelligible, and least understood pathological conditions, and to conclude with those of an opposite character.

Bartlett also speaks in this letter of his acquaintance with Oliver Wendell Holmes, who was to become a kindred spirit of Bartlett in his philosophy of medicine and poetry. He says of Holmes:

> His mind is quick as lightning and sharp as a razor. His conversational powers are absolutely wonderful. His most striking mental peculiarities consist in a power of comprehensive and philosophical generalization, on all subjects—and in a fecundity of illustration that is inexhaustible.

The admiration seems to have been mutual, and the students at Dartmouth seem also to have held Bartlett in admiration. At the end of the session, after Bartlett had left Hanover for Pittsfield, Holmes wrote to Bartlett (B15:129:2) saying:

> The students looked quite inconsolable at your departure and the hotel seemed like a hearse-house, and on the whole you will find it hard to discover a place where you will be more warmly welcomed or more willingly relinquished either by your pupils or your colleagues.

Bartlett's time in Hanover, however, was apparently much less happy than his time in Pittsfield, or, at any rate, other pastures seemed even greener than Dartmouth's. By 1840, Bartlett was actively seeking a new appointment. There are letters from Horace Mann (Y1:4:2) and Benjamin Silliman (Y1:4:3) recommending Bartlett for a position at the College of Physicians and Surgeons in New York. Bartlett was not to go there until near the end of his life, however. Medical schools were beginning to be established in the expanding West and needed professors. The following year, Bartlett wrote to Robert Peter, dean of the medical faculty at Transylvania University in Lexington, Kentucky, first inquiring about the chair in theory and practice of medicine (Y1:5:1) and then expressing interest in the position (R1:5:2).

In 1841, Bartlett was appointed professor of theory and practice of medicine at Transylvania, then the strongest school in the West. Getting to Lexington from New England was, at that time, no small feat. Bartlett describes the journey from Pittsburgh down the Ohio River (R1:5:3). The boat would periodically scrape the river bottom, at which point some passengers were ordered out of the boat in order to refloat it. The boat stuck fast only once, and they needed "tackles and other machinery," which were carried for just such an occasion, to get them afloat once more. Tablecloths were used to soak up leaks in the not-altogether-solid vessel. While Bartlett found the scenery "very beautiful indeed," he describes the towns and villages beyond Philadelphia as "smoky, dingy, dull, odd looking places." Lexington, on the other hand, seems to have met Bartlett's expectations. The class was of a good size, "intelligent, attentive, well behaved" and worth a little over two thousand dollars. Bartlett also seemed pleased with the patients he saw and the fees he received. Money matters surface repeatedly in Bartlett's correspondence. In the "proprietary schools" of the time, lecture fees were the means of support for the professors and most carried on with what medical practice they could manage as well. Bartlett writes to Green in March 1843 (R1:6:1):

> There are a few good families who send for me, and I get occasionally a consultation. We never make a charge less than a dollar; and consultation visits in ordinary cases—the first visit—are $5.00. These few enable me situated as I am, to make even a small and easy business somewhat profitable. I have made one visit twenty-five miles distant, for which the fee was $25; and I saw a second patient, at the same time, incidentally, for $5.00 more. You see from all this, that my place gives me rather more money than I could earn in Lowell, for a much smaller amount of responsibility and labor. I have, hardly, indeed, been called out of bed during the winter. In a business point of view, I feel quite content with my situation if the school can be maintained in its present condition.

Nonetheless, Bartlett expresses doubts about the maintenance of the school in the face of the proliferation of schools and competition from the new school established at Louisville.

It was during his time in Lexington that Bartlett began work on his *Essay on the Philosophy of Medical Science* (H104). His financial success in Lexington and his work on his new book, however, did not seem to prevent Bartlett from becoming homesick. In February 1843, Bartlett was invited to apply for a position at the University of Maryland in Baltimore. Bartlett replied that he would not leave Transylvania without adequate notice, but that he was desirous of returning to the East because his "social relations are so exclusively eastern" (Y1:7:1).

Bartlett was invited to come to Baltimore again the following year (Y1:8:1), and this time he accepted. In April 1844, Bartlett was appointed professor of the theory and practice of medicine at the University of Maryland, although news of Bartlett's appointment apparently slipped out a bit early and against his wishes (R1:6:2; R1:6:3). He took up the professorship in Baltimore in the autumn of 1844. It was during this tenure that Bartlett's *Essay* was published. Despite his move east, Bartlett seems to have continued to feel homesick. A letter from Gilman Kimball in November 1844 (R1:6:4) acknowledges a letter from Bartlett and responds: "I am sorry to know that you are so blue—lonesome—homesick—&c. No wonder if the downfall of Henry Clay and whigging affected you as it has me—I never felt so sad in all my life." Bartlett remained at Maryland for only two sessions.

Also at about this time, Bartlett taught in the medical college at Woodstock, Vermont. He is listed in that school's catalog for the sessions of 1838, holding the chair of pathological anatomy, and from 1845 until 1853, holding the chair of materia medica and obstetrics (Waite, 148-49). The sessions at Woodstock were held roughly from the first of March through June, enabling Bartlett to teach elsewhere during the same years. Bartlett seems to have enjoyed his spring sojourns to Vermont, and to have found leisure time to pursue some new interests. In 1850, he wrote to Augustus Addison Gould, another leading proponent of the philosophy of the Paris school, asking for information and books about fresh water algae (H1).

In the spring of 1845, Bartlett and his wife sailed for Europe. The couple traveled extensively, especially in Italy (Y1:10:1; Y1:10:2). Bartlett also returned to Paris, where he attended medical lectures (Y1:9:1). Two of Bartlett's "Letters from Paris" were published in 1846 in the *Western Lancet*. From these we learn that Bartlett definitely met Louis, and developed what Bartlett calls an "intimate personal acquaintance." (E. Bartlett, "Letter from Paris, May 4th, 1846, 173). Bartlett also attended medical lectures in London (R1:8:2), and describes in great detail his experience at a "magnetic séance."

While Bartlett was traveling in Europe, the faculty at Lexington were arranging for Bartlett's return. In a letter written in February 1846, Robert Peter indicates that the faculty has chosen Bartlett to fill a vacancy and the trustees have confirmed the appointment (Y1:10:3). Peter was apparently concerned at the time because Transylvania had also appointed Samuel Annan, whose 1845 article had criticized Bartlett's *Essay on the Philosophy of Medical Science*. (See also H105.) Peter says that he has encouraged Annan to write to Bartlett to smooth things over, and that he is confident that they will have "at once the most harmonious as well as the most talented corps of professors ever collected in the West." A month later, Peter wrote of his pleasure in knowing that Bartlett was willing to be associated again with Transylvania. Peter wrote again to Bartlett during that summer, when Bartlett had returned to Rhode Island (R1:8:3), saying that he did not think that there was any "danger of a collision" between Annan and Bartlett on the subject of medical philosophy.

Bartlett took up the post in Lexington during the 1846-47 session. By 1848, however, several schools were trying to position themselves to recruit Bartlett's services. Several faculty members at the new University of Louisville were writing to Bartlett to persuade him to leave Lexington for Louisville (Y1:11:1; Y1:11:2; R1:10:3). T. S. Bell (Y1:11:2) intimates that Bartlett was also being wooed to go to New York. The schools at Memphis (R1:10:1) and St. Louis (R1:10:2) also sought to obtain Bartlett's services. It was Louisville that succeeded. On March 2, 1849, J. Cobb wrote to Bartlett from Louisville, saying that the faculty there thought it of the "greatest importance" to be able to announce to their class "even the probable appointment" of Bartlett to their vacant chair. Within two weeks, Bartlett was elected to that position.

In 1849-50, Bartlett took up his position at the University of Louisville, but he remained there for only a year. The University of the City of New York presented an attractive financial opportunity for Bartlett (R1:12:1), and he was offered a position there (R1:12:3) despite the protests of Benjamin Silliman (R1:12:2), who was then Bartlett's colleague at Louisville. Even a student at Louisville wrote to Bartlett (R1:12:4) saying that he was withdrawing from his studies at Louisville, and hoping to come to New York after a year to become Bartlett's private student.

In 1850, Bartlett accepted the chair of institutes and practice in the University of the City of New York. A letter of congratulations came from Charles Dickens just after Christmas that year (R1:13:1). Bartlett's tenure at this school was short, however, for in the spring of 1851, Bartlett was again asked to come to the College of Physicians and Surgeons, then administered by the State of New York, and even at that time a prestigious school. By May, it was apparent that Bartlett was leaving for another position, this time in the same city. Upon the death of Professor J. R. Beck, Bartlett was called to fill the chair of materia medica and medical jurisprudence at the College of Physicians and Surgeons in New York. His appointment came from the regents of the university in Albany on March 5, 1852 (Y1:13:1). Bartlett's *Philosophy of Therapeutics* apparently comes from his work at the College of Physicians and Surgeons. Bartlett lectured there for only two sessions, however, before ill health forced him to retire to his old home in Smithfield, Rhode Island.

Return to Rhode Island

The exact nature of Bartlett's ailment is unknown. A curious sidelight is that Bartlett apparently had complete transposition of his internal organs. That is, his heart was on the right, his liver on the left, etc. (Bauer). This information was originally reported by Sozinskey in 1883. Sozinskey published a letter from Alonzo Clark, a colleague and friend of Bartlett, and the one who edited the posthumous edition of Bartlett's book on fevers. Clark would seem to have been in a position to give reliable information on this. This condition, also known as situs inversus, is associated with bronchiectasis, which Sozinskey refers to as "bronchial catarrh." There is no evidence that Bartlett suffered any ill effects from bronchiectasis or from situs inversus, however, especially given the rather strenuous nature of many of his travels.

Dickson (755) calls Bartlett's terminal illness only an "intractable malady." Burrage (4) reports that Bartlett was afflicted with a "nervous malady," which left him paralyzed but without impairment of his mental faculties. Alonzo Clark calls Bartlett's disease "locomotor ataxia" (Sozinskey, 2). This was a newly described condition at the time and is mostly likely merely descriptive of Bartlett's symptoms; it should not be taken as the tabes dorsalis of late stage syphilis. The knowledge of presently recognized neurological conditions was rudimentary or non-existent in the

1850s and so no diagnosis of Bartlett's condition made at the time should be considered definitive.

The National Cyclopædia of American Biography (70) reports that Bartlett's lingering disease was caused by lead-poisoned water; this was apparently Bartlett's own understanding of his malady. Bartlett's illness was protracted. The people who wrote to him began to mention his illness by 1851, when his friend Roby wrote that "it made us glad to hear that your health had improved since your return home" (R1:14:1). In April 1853, B. R. Palmer invited Bartlett to come to Woodstock to give some lectures. In this letter, he acknowledges that Bartlett's health was getting worse (R1:17:1). In July 1853, C. R. Gilman (R1:17:2) wrote to Bartlett about pains that he was having, and relates Clark's opinion that the pains must be due to lead poisoning. Gilman was planning to have tin pipe put in his house "so that if it be Lead—I'll no more of it." Gilman was hoping that Bartlett might be more willing to come for a visit if he could offer "aplumbic" water.

Bartlett died in Smithfield, Rhode Island on July 19, 1855. The *Boston Medical and Surgical Journal* (1855) reported on Bartlett's death:

> Elisha Bartlett, M. D., died at his residence in Smithfield, R. I., July 18th [*sic*], at the age of 51 years, having long been the victim of a painful neuralgic affection, which compelled him, last fall, to retire from all active employment. Dr. Bartlett was well known as a writer of eminence, being the author of several works on medical subjects of a very high character. He was also a frequent contributor to various scientific and literary periodicals. His most celebrated work is his treatise on typhus and typhoid fevers, which has received the highest commendation from the most eminent medical men. He was professor successively in the Berkshire Medical Institution, the Vermont Medical College, the Transylvania College at Lexington, Ky., the Medical College at Louisville, Ky., and Medical College at Baltimore, and finally, the College of Physicians and Surgeons in New York. In his private capacity he was distinguished for his purity of character, his stern integrity, and his kindness and social virtues.

Oliver Wendell Holmes contributed a lengthier eulogy a few weeks after Bartlett's death. Holmes, also writing in the *Boston Medical and Surgical Journal* (49), said of Bartlett:

> Hardly any American physician was more widely known to his countrymen, or more favorably considered abroad, where his writings had carried his name. His personal graces were known to a less extensive circle of admiring friends, and yet his image is familiar to very many who have received his kind attentions, or listened to his instructions, or been connected with him in the administration of public duties.

Even though he died short of his fifty-first birthday, Elisha Bartlett could claim a wealth of experience about which to philosophize. We turn now to that philosophy, the central topic of this book.

PHILOSOPHY OF MEDICINE IN THE NINETEENTH CENTURY

In order to appreciate Bartlett's philosophy of medicine, it is necessary first to understand the state of American medicine in the first quarter of the nineteenth century, a time of competing theories of disease and competing schools of practitioners. It is also necessary to realize the influence of the Paris clinical school of medicine, which was brought to America by Bartlett and his fellow travelers.

Early Nineteenth-Century American Medicine

When Elisha Bartlett took his medical degree in 1826, the medical world into which he entered was characterized by the presence of many systems, constructed according to the rational ideals of the Enlightenment. These systems were built upon assumptions that were often shared by particular groups of physicians and their patients. In the medical world seen as a whole, however, systems based on widely divergent assumptions conflicted and competed for the allegiance of physicians and patients.

The ultimate roots of these systems are in classical antiquity. Because of the lack of what we would today call empirical knowledge, physicians could do little more than to construct explanatory theories that would be consistent with some deeply held assumptions. Rosenberg (12-13) describes the central nineteenth-century assumption as a "deeply assumed metaphor" in which the body is seen as a "system of dynamic interactions with its environment." Health and disease resulted from the complex and cumulative interaction of the body's own constitution and the circumstances of the environment. Two subsidiary assumptions, according to Rosenberg, influenced how this interaction is organized. First, every part of the body was taken as essentially related to every other part. Hence, local lesions could have systemic consequences and systemic imbalances could result in local ills. Second, the body was seen as a system of "intake and outgo." This bodily system had to remain in balance for health to be maintained. Therapeutics emphasized diet, excretion, perspiration and ventilation in an attempt maintain or restore balance.

Despite this seemingly coherent and shared metaphor, a number of remarkably different systems were developed as particular ways to instantiate the metaphor. The eighteenth and nineteenth centuries were especially fertile for such system building. A few examples will suffice to give an impression of the medical thinking of the time.

In Britain, William Cullen (1710-1790) explained the symptoms of disease on the basis of their physiological causes, remote and proximal, external and internal, predisposing and exciting. He did not attempt to establish etiology, but rather emphasized his physiological principles. For Cullen, almost all human disease was what he called "nervous." Cullen's students developed this idea into their own systems. John Brown (1735-1788), for example, believed that all disease was caused by either an excess or deficiency of what he called "excitability," the capacity to react to external stimuli. When excitability was in proper balance, the result was health. Too much excitability resulted in *sthenic* disease, while too little excitability resulted in *asthenic* disease. Treatment was aimed at restoring balance. For sthenic disease, measures such as bleeding and purgatives, which aimed at depleting excitability, were used. For asthenic disease, stimulants were prescribed. Opium and alcohol were Brown's choices as stimulants. In France, F. J. V. Broussais (1772-1838) taught a system based on a theory of life quite similar to that of Brown. Organisms live by virtue of stimulation or irritation. Disease was the result of normal physiology gone awry. Inflammation was caused by excessive stimulation; hence, treatment consisted of depletive remedies. Broussais favored blood-letting through leeches (Bynum, 45-46).

In America, Benjamin Rush (1745-1813), who was one of the most prominent American physicians of the generation before Bartlett, propounded the theory that formed the heart of the medical orthodoxy into which Bartlett entered and against which he rebelled. Rush's system drew heavily from the system of John Brown and combined the nervous and the hematological systems into one patho-physiological system. For Rush, all disease was caused by an abnormal tension or excitability in the blood vessels. Therapy aimed at restoring health by restoring the proper state of tension. The treatment of choice was massive blood-letting, although Rush also used calomel, a mercury compound used as a purgative (Bynum, 15-18). Because of Rush's long tenure at the University of Pennsylvania, hundreds of new physicians entered practice espousing his system and advocating what became known as "heroic therapy" (Cassedy, *Statistical Thinking*, 53-54).

America in the early nineteenth century was caught in a democratic fervor, which played out in the medical world as a fierce competition of medical systems and practitioners. Starr (30-54) shows that while some physicians were attempting to make themselves into an elite profession with a monopoly of practice, much of the public resisted and insisted on their own rights to manage sickness in their own ways. Three "spheres of practice" coexisted, relatively equal in importance. First, in "domestic medicine," the family was seen as the natural place for most care of the sick. Women assumed the major responsibility and relied upon both oral tradition and manuals written by physicians for domestic use. This practice rested upon the belief that lay people were competent to treat their own diseases. The second sphere, "professional medicine," was largely imported from England, where medicine was a "status profession" and rested upon the monistic theories of disease mentioned above. The British class distinction, however, was not imported; as a result, all sorts of people took up medical practice and began to call themselves doctors. Attempts at licensure were ineffective, and were not strongly associated with graduation from a reputable medical school until much later in the century. Third, the "medical counterculture," or the medicine of the layperson, became a coherently structured rival to the medicine of the physicians. Botanic practitioners, midwives, and natural bonesetters all plied their trades in competition with the physicians. Thomsonian medicine, led by Samuel Thomson (1769-1843), a New Englander with no formal education, was especially prominent. Thomson's system was characteristically simple. The body is composed of four elements: earth, air, fire, and water. All disease has one general cause—cold—and has one general remedy—heat. The way to restore health was to restore heat by clearing the system of all obstructions so that the stomach could digest food and produce heat, or indirectly to restore heat by causing perspiration. Homeopathy was introduced into America in 1825, although it did not gain a significant number of adherents until around 1850. Still, it is another example of a medical system built upon a highly elaborated philosophical doctrine. Homeopaths held that disease was primarily a matter not of physical laws but of spirit. Nearly all diseases were the result of suppressed itch, or "psora." Cures were governed by the "law of similars"—like cures like—and were effected by administering minute dosages of a drug that in larger amounts would cause the symptoms of the disease (Starr, 96-97).

This cacophony of medical voices was in fact a cacophony of medical philosophies. What they shared was a predilection for rational system. This was as much the case for the well educated physicians as it was for the lay person. These systems made sense to their proponents, and probably had some therapeutic effectiveness in certain cases of disease. No nineteenth-century system, however, came close to the effectiveness of the medicine that would come at the end of the century, following Pasteur's discoveries and the acceptance of the bacterial theory of infectious diseases. It was at least partly the spirit of empiricism advocated by Bartlett that could enable such discoveries. Such a medical revolution, however, would demand the abandonment of the spirit of reliance on purely rational system.

The Paris Clinical School

This "spirit of system" was the target of Pierre Louis of the Paris clinical school, which Bartlett came to represent in America. It is in the Paris clinical school that most historians find the origins of modern scientific medicine (Warner, *Against the Spirit*, 3). The teachings of some of its prominent members will serve to illustrate the ways that the Paris school sought to break down the spirit of system. Pierre-Jean-Georges Cabanis (1757-1808) held that the true instruction for young doctors is not received from books, but from the sickbed. For Cabanis, even knowledge of chemistry and physics is not useful to learn medicine; experience is sufficient. René-Théophile-Hyacinthe Laennec (1781-1826), inventor of the stethoscope, held that all medical knowledge comes from observation. Another empiricist, Philippe Pinel (1745-1826), advocated that statistics be used in medicine as much as possible. Pierre-Charles-Alexandre Louis (1787-1872), Bartlett's greatest inspiration, was champion of the "numerical method" and rejected the formulation of hypotheses, holding that true science was but a summary of facts (Ackerknecht, *Medicine at the Paris Hospital*, 3-10). The numerical method was actually little more than a straightforward system of observation, counting and tabulation of symptoms, lesions, disease categorizations and evaluations of therapy. Louis matched patients for diagnosis, age, general condition, etc., and then carefully recorded the effects of treatments that were varied with respect to timing and strength (Bynum, 42-44). This numerical method was enthusiastically adopted by many Americans sympathetic to the Paris clinical school and provided the American medical establishment with a new appearance of scientific respectability (Cassedy, *Statistical Thinking*, 68-91).

The development of the hospitals in Paris enabled the implementation of this empiricist philosophy. Ackerknecht (*Medicine at the Paris Hospital*) argues that the hospitals, along with such technological innovations as the stethoscope and pathological tissue examination, allowed the development of systematic clinical observation and clinico-pathological correlation with autopsy results, and provided the setting for a radically new kind of clinically based instruction in medicine. Michel Foucault (*Birth of the Clinic*) agrees that all these facets in the Parisian hospitals served to establish a new sort of observational laboratory, which he calls the "clinic," but he spins a tale far different from Ackerknecht's story of progress. Foucault sees an objectification of the human body, a turning of the body into a thing to be observed, and a resulting dehumanization of the person by the medical establishment. Many contemporary writers in the philosophy of science recognize that Foucault's story reveals more of the subtleties of discovery in medical science. Social and political realities shape what we take to be scientific facts; medical science is not just a simple and progressive accumulation of facts. No matter how one interprets the significance of the Paris clinical school, however, there can be no doubt that a radical attack on the rational systems of medicine was taking place.

Influence of the Paris Clinical School in America

There was in America a growing awareness of the reputation of the Paris hospitals, driven by the reports of the few Americans who were able to visit them in the very early years of the nineteenth century. The practical hospital experience that was available in Paris was virtually unavailable in the United States, and it eclipsed the opportunities available in London (Warner, *Against the Spirit*, 70). Legal reform in France after the Revolution integrated medical education into a single system. The law provided for an adequate supply of cadavers for anatomy classes from those dying in the hospitals. The hospitals also served as a resource for medical students to learn about disease at the bedside. The "old medicine," overly concerned with theory, was replaced by the "new medicine," devoted to practice (Bynum, 28). By the mid-1820s, with the end of the Napoleonic Wars, Paris became readily accessible to American travelers, and American students and recent medical graduates began to visit in greater numbers (Warner, *Against the Spirit*, 33-39). Elisha Bartlett was one of them.

Bartlett exemplifies the devotion to Pierre Louis of the Americans who went to Paris (Warner, *Against the Spirit*, 167-169). Osler ("The Influence," 199) lists thirty-seven American students who followed Bartlett to Paris between 1830 and 1840. This list includes many of the most prominent names in nineteenth-century American medicine, and many whose reviews and letters are cited in this book. Even though many of these men apparently had little, if any, personal contact with Louis, they brought back to America a deep devotion to Louis, a devotion that Oliver Wendell Holmes ("Some of My Early Teachers, 431) saw as tending almost toward idolatry.

It should not be surprising that those returning from Paris would meet resentment from the American physicians who remained behind. They were, after all, returning with an iconoclastic medical philosophy that disparaged the predominant theoretical systems and brought a skepticism towards prevailing methods of therapy (Warner, *Against the Spirit*, 148). Many of those returning from Paris, however, applied this skepticism even toward French therapeutics, which they found not completely applicable to the particularities of the American context. In addition, many of these returning Americans, while applauding the opportunities that the Paris hospitals provided for learning and scientific study, condemned the objectification of patients in the realm of therapeutics (Warner, *Against the Spirit*, 253-290). Although they recognized the superiority of therapeutics in London, they chose to go to Paris to gain knowledge and experience in areas other than therapeutics, especially in diagnostics and pathological anatomy, and this experience was much easier to obtain in Paris than in the more restrictive organization of the London hospitals (Warner, *The Therapeutic Perspective*, 198).

The most fundamental of the tenets of the Paris school was its insistence on empiricism as the way to the truth and its rejection of rational systems. Warner (*Against the Spirit*, 165-67) argues that most antebellum American intellectuals made frequent references to Baconianism and were committed to an empiricism at least in the spirit of Francis Bacon. The physicians, however, put a French twist on this Baconianism. In doing this, they could link their creed to their personal experience with Louis and others of the Paris clinical school in a way that they could not do directly with Bacon, and could at the same time proclaim a post-1812 polemic against Britain.

As King (*Transformations*, 183-85) points out, in most of the eighteenth century, empiricism was a pejorative term, implying ignorance of medical theory; but by the early nineteenth century this was changing. The growing spirit of experimentation and analysis gave increased value to experience. It rejected the purely rationalistic methodology of the eighteenth century.

While this empiricist philosophy does not seem at all sophisticated to the academic philosopher and at times looks hopelessly naïve even when judged by the debates of eighteenth-century philosophy, it is, nonetheless, an important development in philosophical reflection on medical knowledge and practice. For even if the empiricism of the Paris clinical school is thought to be philosophically naïve, the rational medical philosophy of the systems must be even more so, for those systems are more akin to the sort of medieval metaphysical speculation that Descartes and virtually all who followed him repudiated.

ELISHA BARTLETT'S PHILOSOPHY OF MEDICINE

The philosophical position espoused by the Paris school was not unique in the nineteenth century. In fact, many early nineteenth-century thinkers, including Thomas Jefferson, were denying any general explanatory function to science. Science was, according to them, restricted to the description and collection of facts and the classification of facts. This was in large part a reaction to the overly ambitious construction of rational systems in the eighteenth century. Even Jefferson advocated the abandonment of hypothesis for "sober facts" (Daniels, 343-44).

Bartlett's *Essay on the Philosophy of Medical Science* provides what is generally acknowledged as the most complete American account of the philosophy championed by the Paris school (Warner, *Against the Spirit*, 175; Cassedy, *Statistical Thinking*, 66-67). Huntington (13) calls the *Essay* "the work on which Dr. Bartlett's fame, as an author, will mainly rest, and by which his relative place, among the medical writers of our day, will be established." He remarks (15): "Some may dissent from the author's reasonings and conclusions, but all must admit the fairness and ability with which he has conducted the discussion." Osler ("Rhode Island Philosopher," 133) goes so far as to call this work "a classic in American medical literature." Erwin Ackerknecht ("Elisha Bartlett," 43) says that Bartlett's *Essay* "might turn out to be an important document transcending the limits of American medical history" and that it "allows us to study in toto the philosophy that formed the basis of one of the most progressive movements in medicine, its strength—and its limitations" (60).

The *Essay* is Bartlett's most complete statement of his philosophy of medicine. For Bartlett, phenomena are not the materials out of which science is fashioned according to some theory; the phenomena are themselves the science. Hypothesis and theory are nothing more than speculation. The phenomena we observe are all we know, and all we can know. Bartlett's philosophical endeavors, however, go beyond this one seminal work. Indeed, if the *Essay* is all that one reads, one is reasonably left with a sense that further explanation is warranted on several fronts. To set the *Essay* in the larger context of Bartlett's philosophy of medicine, one should consider three other works: *An Introductory Lecture on the Objects and Nature of Medical Science*; *An Inquiry into the Degree of Certainty of Medicine, and into the Nature and Extent of Its Power over Disease*; and *The Philosophy of Therapeutics*.

An Introductory Lecture on the Objects and Nature of Medical Science

An Introductory Lecture on the Objects and Nature of Medical Science was delivered by Bartlett at Transylvania University on November 3, 1841, at the beginning of his tenure in Lexington. It is a good introduction to his medical epistemology. Bartlett says that his object is "to ascertain the essential and true character of medical science; to find out in what it consists; what are its elements, what are its objects of investigation, and what the true methods are by which we can attain them." In other words, "what is it that we wish to know? what is it that we can know? and what are the true and best means of arriving at this knowledge?" (4). Just as with chemistry and physics, the object of medical science is the ascertainment of facts; it is only the unique subject matter, the natural history of disease and methods of cure, that makes medicine different. According to Bartlett, medicine's "legitimate object" is "the investigation and ascertainment of all the phenomena of morbid action—the relations of these phenomena to each other, and to their causes—and, also, to those substances and agents in nature which are endowed with this property of influencing and modifying them" (6). This doctrine is not generally recognized by physicians, claims Bartlett, and this leads them to speculations "that can only be characterized by the terms metaphysical or transcendental." These theorists go wrong by directing their inquiry not toward the phenomena of morbid action and their ascertainable causes, but rather toward "their intimate, ultimate, and essential nature," which consists of "subtle and inscrutable processes" (7). Bartlett claims that true medical science aims, through purely empirical means, to arrive at principles, which are simply the "universality of the fact," ascertained by a sufficient number of observations (8). Bartlett recognizes that observation in medicine cannot produce certainty. The reason that medicine is less certain than other sciences such as chemistry and physics lies in the imperfection of our methods of observation. Bartlett claims: "The ultimate laws and principles, connected with, and arising from the vital forces, and their relations, are just as absolute and immutable, as those connected with the sciences." The difficulty is that in medicine, the phenomena and their relations themselves are so complex, making the process of observation difficult and conclusions uncertain (9). When observations are made on a vast scale, however, this uncertainty is lessened (11). Bartlett says that he has no objection to theory or hypothesis in medicine, as long as they are kept in their proper places, which, in medicine, are "very subordinate and very humble ones." Theory is just "explanation and interpretation" of phenomena (12-13). Theories are assumptions and not phenomena, and are not the proper matter of science. Bartlett claims to be a "humble but earnest disciple" of Lord Bacon's philosophy, and exhorts medical science to a "fuller recognition and practice of the true, simple, and rigorous laws of the Baconian philosophy" (15).

One can see in this address a prefiguring of the major theses of Bartlett's *Essay*. A notice of the *Introductory Lecture* in the *Boston Medical and Surgical Journal* for January 1842 admires Bartlett as "a candid medical philosopher and a gentleman," but the writer also seems to intimate that Bartlett is here tackling insoluble problems: "Though we by no means wish to alarm pathological peace-makers, we must be allowed to say that a medical millennium is still in the obscurity of the future." Part of the obscurity here lies in an explicitly stated but unargued metaphysical claim about the regularity of the laws governing the functions of the human body. Bartlett would return to this point in his later writings.

An Essay on the Philosophy of Medical Science

The *Essay* is divided into two parts: "The Philosophy of Physical Science" and "The Philosophy of Medical Science." Each part is headed by several "propositions," which summarize Bartlett's major points. The chapters then elaborate on these propositions.

In the first part, Bartlett discusses six propositions concerning science in general. These propositions are fine summary statements of the empiricist philosophy that Bartlett wanted to apply to medicine. The first is that "all physical science consists in ascertained facts, or phenomena, or events; with their relations to other facts, or phenomena, or events; the whole classified and arranged." Second, the facts and relations that constitute science can be ascertained in only one way, and that is through observation, or experience. Facts cannot be deduced or inferred from other facts. Third, a law or principle of physical science consists in a "rigorous, and absolute generalization of facts, phenomena, events, and relationships." It is identical with "the universality of a phenomenon, or the invariableness of a relationship." Fourth, hypotheses are attempted explanations of phenomena and relationships. They are assumptions. Science itself is independent of hypotheses.

Fifth, theory is one of two things. It is either a generalization of phenomena and relationships, and hence identical to a law, or an attempted explanation of phenomena or relationships, and hence identical to a hypothesis. Sixth, all classification or arrangement "depends upon, and consists in, the identity, or similarity, amongst themselves, of certain groups of phenomena, or relationships."

In the second part, Bartlett puts forth five propositions specific to medical science. The dependence upon what he has previously said about all science is obvious. First, all medical science consists in ascertained facts and relations, classified and arranged. Second, each separate class of facts can be ascertained in only one way—by observation and experience. Classes of facts cannot be deduced or inferred from other facts. Neither etiology nor therapeutics, for example, can be deduced from pathology. Third, a law of medical science is an "absolute and rigorous generalization of some of the facts, phenomena, events, or relationships, by the sum of which the science is constituted." The actual laws of medical science are, "for the most part, not absolute but approximate." This is because they are based on the numerical method, and hence dependent upon the calculation of probabilities. Fourth, most medical doctrines are hypothetical explanations or interpretations of ascertained phenomena. Hence, they do not constitute a "legitimate element" of medical science. Fifth, diseases can be classified and arranged, and such classifications will be "natural and perfect" in proportion to the degree of similarity between the diseases themselves.

Bartlett sensed the controversy that he would provoke, even while he was at work on his *Essay* during the winter of 1843-44. In February 1844, he wrote to Holmes from Lexington (H104):

> I have been at work, for a good part of the winter, in the preparation of a little book, which I shall call "An Essay on the Philosophy of Medical Science" with this motto, from some old cock, quoted by Whewell,—I forget his name just now,—"I trust that I have got hold of my pitcher by the right handle." I have attempted a pretty full, and elaborate development of what I conceive to be the true nature and philosophy of medical science,—so generally, according to my notions, either very partially and loosely apprehended, or altogether and totally misapprehended, and misunderstood. I have endeavored to vindicate and restore the empirical philosophy in all its purity and absoluteness; and to exhibit the true character of all dogmatism, rationalism, and à prioriism.—the evil genii of our science. My doctrines, I am well aware, will encounter,—if they are considered of importance enough to attract any considerable degree of notice,—stiff opposition; and I shall look to you as one of the few men disposed and qualified to stand by them.

Reaction to Bartlett's book came swiftly and forcefully in the medical literature. Holmes, naturally, was sympathetic to the ideas of the Paris school that Bartlett was advocating. Bartlett wrote from Baltimore, where he had moved by the time the book was published, to Holmes in November 1844 (H105):

> I wish to thank you more formally and emphatically than I have yet done for the more than kind, for the enthusiastic, reception which you have given to my little book. It was the white day of my scientific life, the culminating point of my literary orbit—that on which I received your own letter and the one from Dr. Jackson—they both gratified me beyond the power of expression. I had looked, especially and particularly to you two, for an approval of my book, but I had not ventured to anticipate so warm and so hearty a

> welcome. Will you convey to Dr. Jackson the thanks which I send to you? Your letters will help me to look quietly on, at the stupid and gaping wonder with which the blockheads of the profession will greet my book. The Philadelphia Examiner has already opened the brainless halloo and I look for it to be followed up, in full chorus, both at home and abroad. Dr. Annan made it the subject of his Introductory at the Washington college in this city;—asking his hearers, as I am told, with a due holy, orthodox, scientific horror, how they could seek the instructions of a teacher, who avowed himself to be an empiric! As to reviews, I have done nothing. I shall leave it in the hands of its friends and its foes. It would gratify me to have an appreciating notice of it—such as Dr. Jackson or yourself would write—in the N. American. Parker has it for the N.Y. Journal. It does not lie much in his line of thought, but he likes it very much, and will act accordingly.

Two months later, Bartlett wrote to Holmes again (R1:7:1), asking him to write a review of the book. Bartlett mentions a review that Dr. Shattuck had written, but that was unsuitable for publication. Roby (R1:7:2) reports that the review was too long, according to the editor, but that there were probably other reasons that the review would not appear.

Despite all this behind-the-scenes maneuvering, it does not seem that the reviews to which these letters refer ever were published. Nevertheless, several reviews and notices of Bartlett's *Essay* did appear shortly after its publication. Two of the more detailed of these were in the *Southern Literary Messenger*. The first, by "J. S. A.," who is identified in the *Boston Medical and Surgical Journal*'s notices as James S. Allan, appeared in June 1845. Allan's review is more a panegyric on Lord Bacon's philosophy, calling it "the crowning achievement of human genius," than a review of Bartlett's *Essay*. Still, it does manage to praise Bartlett as well, calling his work "the clearest interpretation of the Inductive philosophy that we have met with" (331). Allan condemns those who would corrupt the pure Baconian philosophy, especially those "speculators," the German transcendentalists, by "superadding to [Bacon's] method the idealism of Plato and the empty dialectics of Aristotle" (330). He applauds Bartlett's suggestion that we should drop the term "inductive reasoning," and expounds upon the vagueness of the term, which was being used almost synonymously with "reasoning." He claims that Bartlett has "beaten all competitors in simplifying the philosophy of science" (335). Allan considers two objections against Bartlett's philosophy (338). First is the claim that such a strict empiricism would paralyze physicians from rendering any treatment or making any advances "when no longer lured by the extravagant hopes and ideal charms of hypothesis." Allan answers that the human tendency toward speculation is much too strong ever to allow empiricism "such an absolute sway as would be requisite before it could begin to manifest the peculiar abuses and excesses toward which it may have a leaning." This is a very peculiar argument itself, for it seems to defeat the ultimate correctness of the empiricism that it seeks to defend. The second objection to Bartlett's philosophy has to do with its rejection of hypothesis. Allan rightly answers that Bartlett does not reject all hypothesis, but merely makes hypothesis subordinate to his science of observation. It should be kept in mind that Bartlett tends to use "hypothesis" in a rather loose sense, meaning an explanatory conjecture. Bartlett would likely not object to a present-day use of hypothesis as something to be verified empirically.

Samuel Annan's review, which prompted the above-mentioned concern about uniting Annan and Bartlett on the same faculty, was far less laudatory. The *Boston Medical and Surgical Journal* for October 29, 1845, says that Annan criticizes the work "as though he owed it a grudge," and finds it fortunate for Annan that Bartlett was at the time traveling in Europe, allowing the reviewer to "thrash the shadow with perfect impunity." It predicts a "literary retaliation" from Bartlett.

Annan presents a systematic attack on Bartlett's propositions. He finds Bartlett's criticisms of the present state of physical science to be "gratuitous and misplaced." One of Bartlett's "cardinal errors," according to Annan, is that he does not properly distinguish between theory and hypothesis. Annan maintains that theory is nothing more than a concise statement of what we have learned through observation and experiment, whereas hypothesis is a rational conjecture, and is only an attempt at an explanation when the established facts are insufficient to establish a theory. Annan refutes the idea that scientists are satisfied with establishing hypotheses without subjecting them to experimental observation (610). He then argues that Bartlett does not adequately specify how the classification and arrangement of facts is to be carried out and how such an arrangement and classification is to count as knowledge. Annan would require the construction of theory to explain how to carry out and interpret the arrangement and classification. Next, Annan accuses Bartlett of making a straw man in his assertion that some would ascertain fact in some way other than though observation. Scientists, according to Annan, always first observe and then reason about their observations. Annan was favorably disposed toward Benjamin Rush, who was one of Bartlett's prime targets, especially concerning his theory of fever. Annan derides Bartlett for his criticisms of Rush and Cullen, well established regular physicians, and for grouping such physicians with the likes of Hahnemann and Thomson, whom Annan regarded as quacks. Here, however, Annan seems to rely more on personal reputation than substance, for Bartlett's attack was on rational theory building in general, and not just upon particular theories. With regard to Bartlett's claims that one branch of knowledge cannot be derived from another, pathology from physiology, for example, Annan maintains that this is a mere truism (615). In terms of therapy, Annan maintains the superiority of a rational method over an empirical one. Treatment can be arrived at by reasoning from pathology (617). Empiricism amounts to little more than blind trial and error. Annan ends by denying that mathematical demonstration will ever be of much use for medicine because of medicine's nature as an inexact science.

Although these two essays take utterly opposed positions on Bartlett's *Essay*, they share a penchant for impassioned rhetoric. One wonders whether too much is being made of professional alliances and *ad hominem* argument. Annan, however, does make a good point in demonstrating Bartlett's tendency to use the term "hypothesis" in a way that is inadequately defined and delineated.

In a more purely philosophical review that appeared in the *New York Journal of Medicine and the Collateral Sciences*, a review almost apologetic for its disagreements with Bartlett, Charles A. Lee comments upon Bartlett's failure to show how merely observing phenomena and classifying them can result in knowledge. He rightly observes that Newton's observation of an apple falling from a tree resulted in no knowledge until the pure process of mathematical reasoning was put to work to establish a principle (66). As Lee points out (68), Bartlett even seems to concede as much in his discussion of Fresnel. Lee admits that there are many "false facts" in medicine, and much false reasoning. But this is not to say that there are no dependable facts and no reasoning that is good for anything (69). With respect to Bartlett's claims about deducing one science from another, Lee reasonably claims that knowledge of pathological anatomy does seem to depend upon knowledge of normal anatomy (70). His most pointed criticism, however, concerns Bartlett's claim that therapeutics is not founded upon pathology. Lee argues that reasoning, apart from all experience whatsoever, would lead us to conclude that removal of the cause of a disease would cure it (73). Bartlett is mistaken, according to Lee, in his claim that if two cases of disease are alike, the effects of a given remedy must be the same. Bartlett makes the uncertainties of medicine dependent upon our imperfection in diagnosis and incomplete understanding of pathology. It may be the case, however, that individuals have different constitutions and react differently to a given remedy (74-75). This seems to be a serious and reasonable philosophical difference about the ontology of medical laws. Still, Lee finds much to admire among the faults that he points out in Bartlett's *Essay*.

Another ontological disagreement shows the continuing disdain of some in medicine for the Cartesian and generally modern view that the body's working can be understood according to mechanical models. A short review in the *Medical Examiner* for October 19, 1844, says that the "great and pervading error" of Bartlett's book is "regarding the living body in the light of a mere physical machine." The author takes Bartlett to be dragging forth a centuries-old philosophy, which all around him have proclaimed dead and that ought to remain buried.

In a more purely epistemological vein, L. M. Lawson, in the *Western Lancet* of December 1844 concedes that speculation has done much to harm the utility of medical science; but he argues that reasoning from particulars to general principles is "indispensable to its perfection" and that more error results from relying too exclusively on plain fact without proper explanation (387-88).

Picking up on Annan's critique, the *British and Foreign Medical Review* also takes Bartlett to task for the way he uses some common terms. In particular, Bartlett confuses *scientific* knowledge with *empirical* knowledge. Empirical laws can express only probabilities, while scientific laws predict with absolute certainty because they explain *why* (142-43). With regard to Bartlett's claims about hypothesis, the author accuses Bartlett of forgetting that most accepted scientific facts began as mere hypotheses; hence, the boundary between hypothesis and law is not as clear as it seems at first. Some hypothesis is absolutely necessary to establish fact. One should, however, be ready to abandon a hypothesis that proves to be "unstable" upon empirical testing (143-44). Finally, the reviewer argues that no collection of facts, such as that which Bartlett considers to comprise science, is sufficient to enable one to make predictions of any unknown result from any set of data. Some common principle connecting the phenomena is necessary (145). Again, Bartlett would probably not disagree with the use of hypothesis when the term is used in this way. This critique does raise another substantial philosophical issue, however. What do we take to be the nature of science? Does true science require explanatory theory that describes a real state of the world, or is it merely a collection of facts established with some degree of certainty that will allow empirical success in predictions? Such questions continue to be controversial to this day.

The *Buffalo Medical Journal and Monthly Review* waited two years before publishing a review of Bartlett's *Essay*. This review takes up several of the ideas already put forth. The reviewer disputes the claim that there is no difference between the phenomena of life and the phenomena of physics (12-14). He also disputes the claim that reasoning has little to do with the building up of science (15-16). Mere observation, without reasoning, does not yield knowledge. This author also notes that hypotheses are necessary; only when they withstand empirical testing do we gain scientific knowledge (16). He also disputes Bartlett's claim that we can understand pathology without any knowledge of physiology (18). Empiricism and rationalism should not be set against one another in the realm of therapeutics. The science if therapeutics requires the interaction of both. Finally, the author disputes the importance of the numerical method in medicine (20-22). Its application is limited to those diseases that are susceptible of ready and clear diagnoses, and medicine is inherently imprecise. Medical practice involves exceptions as much as regularities. The numerical method ought to be used, but should not supersede rational investigation. This claim seems to rest on an ontological assumption that the workings of the body are different in kind from the workings of the rest of nature.

Josiah C. Nott, another student of Louis, writing in the *New Orleans Medical and Surgical Journal*, seems to appreciate more subtleties in Bartlett's arguments. He calls Bartlett's *Essay* "the most remarkable medical book yet written in this country," and says that it "tears off the veil which has been thrown over false science, and exposes all its deformities" (491). Nott observes that Bartlett is not disposed to discard theories and hypothesis entirely from science, but only insists that theory and hypothesis always be "dealt with rigorously *as such*, and never stated as established laws. In this way, they may be useful in stimulating or directing investigation; but otherwise they may lead to mischievous consequences" (494).

In 1853, Edwin Leigh wrote a Boylston Prize winning essay entitled "The Philosophy of Medical Science" and subtitled it with "special reference" to Bartlett's *Essay*. Leigh's essay is probably the most philosophically sophisticated of the critiques of Bartlett's own day. Leigh insightfully remarks that even the most earnest empiricists who will not admit to any "philosophy *in* science" will insist upon "their peculiar philosophy *of* science" (3). No one can study science or medicine without *some* philosophy. Articulating this philosophy is, of course, just what Bartlett was doing. Leigh admires what Bartlett is trying to do, but doubts whether his philosophy is the correct one. Science cannot be only "observed facts." Leigh argues that Bartlett's philosophy is not based upon observation; hence, it amounts only to the sort of speculation that he is trying to be rid of (6). There must be some fundamental principles that *allow* the observation of phenomena in the first place (9-10).

Leigh further criticizes Bartlett for arguing that because some facts cannot be proved from other facts, therefore no facts can be proved from other facts. This is clearly fallacious reasoning. Leigh admits that all particular facts ought to be confirmed by observation. But this does not mean that there are never sufficient reasons for arguing from one class of facts to another. For example, by our knowledge of anatomy of the urinary system, we know that the kidney must secrete urine. Yet the kidney actually secreting urine had not at that time been observed. Hence, the fact must have been established through reasoning (13-14). Therapeutics depends largely upon reasoning from cause to effect (15). Reasoning by exclusion results in facts, but these are not observed facts (15). Laws of uniformity and laws of causation, to which Bartlett refers on pages 79 and 87 of his *Essay,* are not arrived at by observation.

The great vice of Bartlett's theory, according to Leigh, is that it excludes from science all its "ideas, thought, truths and principles, leaving nothing but an array of lifeless material facts" (15). Leigh plausibly argues that "a law is a general truth proved by the facts, and not a general phenomenon observed in the facts." It is not, as Bartlett (148, 175, 220) says, the phenomena and the relationships themselves (16). General truths, for Leigh (16-17), are deduced from particular facts. For example, "vertebrae" is a general concept arising from the observation of many like bones. This seems to be just what Bartlett has in mind when he talks about arranging and classifying. Yet Leigh argues that laws comprise another ontological level and are not simply the phenomena themselves, arranged and classified.

Leigh admits that we must rid ourselves of as many previously-formed opinions as possible so that we can be impartial observers free of scientific prejudice (18). But it is impossible to observe the phenomena of nature with no ideas about what we might expect to find. The human mind just cannot do this. Leigh thus nods to Kant's categories.

Bartlett alludes to the imperfection of science. But are such imperfections removed by eliminating hypothesis and reasoning from medical science? According to Leigh, most error comes not from faulty reasoning, but from imperfect observation. The only remedy for this is "sound judgment and clear discrimination in the observer" (21). Science should drive away "ridiculous theories, absurd hypotheses, and false doctrines" but Bartlett has also driven away "rational hypotheses" which are like partial truths from which we discover the complete truth (23).

While present-day historians such as Warner and Cassedy tend to recognize Bartlett's *Essay* mainly as an important articulation of an important movement in nineteenth-century American medicine, Lester King has vigorously criticized Bartlett's philosophy, and King does make several good points. First, Bartlett never clearly distinguishes fact from phenomenon. Are facts, phenomena, and relationships synonymous, and are they all equally a matter of observation? As King points out, Bartlett understood "fact" as resting on a succession of individual experiences. Such a notion of fact, however, requires an element of abstraction. To use King's example, "Fire burns" is a fact, but it is also an abstraction or generalization that links a whole set of other facts, laws and predictions (*Medical*

Thinking, 250-51). In equating facts and phenomena and in holding that science is nothing more than an organized accumulation of facts, Bartlett did not seem to appreciate the difference between an observation, recorded as a datum of experience, and a generalization (*Transformations*, 205). Even to this day, the notion of fact is philosophically controversial; even if the ambiguities in Bartlett's account were clarified, there would still be no agreement about the nature of a fact (Stempsey, ch. 3).

King ("Medical Philosophy," 156-59) accuses Bartlett of making the same errors, or, at any rate, errors comparable to the eighteenth-century attitudes he attempted to quash. Bartlett often used words such as "absolute," "indisputable," "always," and "never." King sees these as signs of "uncompromising dogmatism" and quite incompatible with empiricism. He finds nothing original in Bartlett's thought and describes him only as "an eddy in the stream of progress." King argues that other nineteenth-century authors have given better accounts of the philosophy of medicine. Even if this is true, however, King may be overly harsh in his disparagement of the *Essay*, for Bartlett's book is still generally recognized as the best example of the thought of the Paris clinical school presented by an American.

King also claims that Bartlett had no real awareness of the methodology of the "more exact sciences," even though he claimed that medical science followed them. His refusal to admit hypothesis into the realm of science illustrates this. King gives an example from Bartlett's own time. Irregularities in the orbit of Uranus prompted astronomers to hypothesize the existence of another planet. In 1846, after a relatively brief search, the planet now called Neptune was found. King suggests that Bartlett's motivation for all his excesses was a simple hatred of the spirit of "system" that continued to rule in medicine (King, *Transformations*, 206-7).

In sharp contrast, Erwin Ackerknecht ("Elisha Bartlett," 59-60) appreciates Bartlett's philosophical treatise as a systematization of the ideas of the major figures of the Paris clinical school, even if they are not original, and are "often to the point of literal agreement of formulations." Ackerknecht opines that since Bartlett was trained in a methodology where research could be carried on only in hospitals, but "condemned to the position of an itinerant teacher and practitioner," his systematic *Essay* was the only way he could advocate the new French philosophy and make an original contribution "to clean away the rubbish of the past."

William Bean also portrays Bartlett more sympathetically, as having "attributes of gentleness, grace, dignity, and strong character," and taking aim at "the fabulously lethal influence of Benjamin Rush" (321). Bean claims that Bartlett was able to see nature in an unbiased way and knew that bleeding people with yellow fever harmed or even killed them. Rush must have observed this, but he remained determined to clear out the poison by bleeding. Such was the hold of system on him (325). We may question whether any observer can be so completely free of bias and observe nature "as it is." Nonetheless, Bartlett's philosophy, even with all its limitations and sometime lack of subtlety, is valuable as a systematic critique of a sometimes deadly myopia in the eighteenth-century "spirit of system."

An Inquiry into the Degree of Certainty of Medicine, and into the Nature and Extent of Its Power over Disease

In 1848, Bartlett revisited the realm of the philosophy of medicine with the publication of an eighty-four page monograph entitled *An Inquiry into the Degree of Certainty of Medicine, and into the Nature and Extent of Its Power over Disease.* This fundamental question of epistemology was addressed by the French writer Cabanis before him, so it is not surprising that Bartlett should tackle this topic. Bartlett here argues that medicine does deserve praise for its advances in several areas. He takes anatomy as a prime example of a science where certainty is indubitable. He does not claim that knowledge of anatomy is complete, but he does express the optimistic view that eventually all that is knowable will be known (16). The certainty of pathology and therapeutics, however, is less well established. Bartlett uses the example of pneumonia to illustrate his position. The physical findings of pneumonia are quite certain. The prognosis, however, can be known only with various degrees of probability, and the most uncertainty arises in considering the causes of the disease. With respect to treatment for pneumonia, Bartlett cites several empirical studies and argues that blood-letting is a beneficial treatment for pneumonia, a conclusion that had been reached several years before by Pierre Louis, applying the numerical method to his observations (Warner, *Therapeutic Perspective*, 203-4).

Bartlett makes an even stronger claim, though, when he says that this fact has its "foundation in nature and in truth" (42). Despite the more general uncertainty in therapeutics, there are a few certainties to be gleaned from studies of the treatment of pneumonia. First, many cases of pneumonia terminate naturally and spontaneously. The exact proportion of cases that will so terminate, however, is impossible to say (50). Second, it is certain that some cases of pneumonia will terminate fatally, no matter what treatment is given. Again, certain probabilities can be specified, e.g., that the elderly and chronically debilitated are more likely to die of pneumonia, but no certainty can be attained (51-52). Third, there is a class of patients in which there is no means of knowing whether the pneumonia will terminate fatally or not. This is not to say, however, that we cannot know with relative certainty that a certain percentage of such patients will die of pneumonia (53). From all this, Bartlett comes to two conclusions. First, with regard to pneumonia the science of medicine, although still "unfinished and progressive," is "to a very satisfactory extent, settled and positive." Second, medical art, principally blood-letting and antimonials, although not completely powerful in treating pneumonia, is still of "great and unquestionable utility" (55).

Next, Bartlett groups diseases into five categories. Unlike many of the nosologists who had preceded him, Bartlett did not make his categories depend upon any rational system; he simply classified diseases according to observed severity and therapeutic effectiveness. The first class consists of mild diseases that rarely threaten life and almost always terminate with a return to health within a few days. The examples he gives are common catarrh and simple acute diarrhea. The second consists of more serious diseases, such as functional dyspepsia, chorea and chlorosis, that rarely destroy life and may be relieved by removal of their causes, although not with such a strong correlation as in the first group. Third is a group of more severe diseases such as sporadic dysentery, simple acute rheumatism, acute pleurisy and tonsillitis, which only sometimes can be adequately controlled. The fourth group is diverse, consisting of the general fevers—continued, periodical, and exanthematous—and hooping-cough [*sic*], Asiatic cholera, erysipelas, and delirium tremens. Most of these, Bartlett says, are self-limited, but might result in either death or recovery. The fifth group are "little or not at all" controllable by remedies, and result in death. Hydrophobia, epilepsy, traumatic tetanus, cancer, and diabetes are examples. A few other diseases, such as syphilis, scrofula, and some chronic skin afflictions, have no tendency to spontaneous cure, but can be controlled by medicine (61-73).

Bartlett concludes his *Inquiry* with a short discussion of medical doctrines. In particular, he considers homeopathy. He criticizes the claims of homeopathy, arguing that many of the cures claimed by homeopathy were in fact cases in which the disease would have spontaneously resolved. Bartlett argues that the rigorous empirical methods of Louis and others are necessary to substantiate any claims of medical knowledge, and that the homeopaths have failed to employ such methods.

Reviews of the *Inquiry* were mixed. The *Boston Medical and Surgical Journal* calls Bartlett "a fearless advocate for common sense in medicine, discarding all vulgar associations and the opinions and whims of speculators in health," aiming to "open the leaden eyes of the world to the monstrous impositions forced into notoriety under high-sounding name, and with the false pretense of being improvements."

In the *Western Lancet*, Lawson, who had previously been critical of Bartlett's *Essay*, says that this work brings him pleasure that Bartlett "should forsake his untenable positions, and now stand forth a firm and unwavering advocate of the power of truthfulness of medical science." But it is hard to see how this assertion is anything more than approval of the rhetoric of some of Bartlett's claims about the obvious certainty of some observations. While Bartlett does expound upon the matter of certainty, his empirical philosophy is unchanged from what he wrote in his *Essay*.

The *Medical Examiner* of Philadelphia is much less kind. The reviewer calls Bartlett's work a "curious production, the like of which we have seldom seen from the pen of any one who had passed the age of a sophomore." He makes much ado about Bartlett's admission that not all cases of pneumonia have predictably certain outcomes, concluding: "So it would appear, from the author's shewing, that the certainty of medicine consists in its uncertainty!" It might appear that Bartlett is doing little more than asserting a truism, but he is in fact basing his conclusion on his observation—that we can indeed have different degrees of certainty about the outcome of some diseases on the basis of our observations of the natural history of the diseases and the results of different means of treatment.

Albert Stillé's review in the *American Journal of the Medical Sciences* is more detailed and thoughtful. He writes of the mood of skepticism prevailing in the

America of his day. A fellow supporter of the Paris school, Stillé applauds Bartlett's use of the numerical method to show that there is a degree of certainty in the diagnosis and treatment of pneumonia, including the effectiveness of bleeding. He also argues that Bartlett's nosology is successful in showing that his claims about certainty extend beyond his carefully considered example of pneumonia. Regarding surgery, that art is no more inherently certain than the art of medicine. Such attributions of certainty to surgery by the unschooled result from seeing surgery only in terms of "its mechanical processes, which appeal so directly and forcibly to the senses of the unskilled."

J. H. Shearman, writing in the *New York Journal of Medicine*, sees Bartlett's work as a close investigation of the important question, "What is our practice worth?" Shearman draws what he calls "a very different conclusion from our author, but not adverse to him." He rightly observes that Bartlett's claims about knowledge are an account of the *effects* of disease, and not of the *causes*. Certainty can be obtained about symptoms and appearances, but not about how and why they are produced. Shearman is not disappointed, then, when Bartlett gives only statistical accounts of the results of bleeding and giving antimony in cases of pneumonia; but he argues that this data should not be considered proof of any positive knowledge about the nature of the disease, its causes, and the mechanisms of action of its treatments. Bartlett would not disagree, for in his *Philosophy of Therapeutics* he seems resigned that ultimate causes must forever be unknown. Nonetheless, this seems to be exactly what Shearman demands in true knowledge, for he says (87-88):

> For an art to claim the respect of a science, by putting forth a statement that after the elapse of two thousand years, the best discovered mode of treating pneumonia, is by bleeding and giving large doses of tartar emetic; neither of which, nor both combined, can do more than mitigate the symptoms in a proportion of cases, is not to make out a strong case. The philosopher and philanthropist will scarcely recognise the claim, and a court of equity would hardly decree in favor of it if contested.

This points out what is perhaps a major bit of pessimism that characterizes Bartlett's writings on the philosophy of medicine: a thoroughgoing skepticism about the possibility of knowing ultimate causes of disease. What counts as *ultimate*, is, of course, a matter of philosophical debate to this day. Nonetheless, Shearman is much more optimistic than Bartlett when he writes (90):

> Let a few choice and noble spirits dedicate themselves to the examination of the elements and organization of our bodies, their functions, and the effects of substances upon them. Let us not be dismayed by obstacles. We possess the Rosetta stone of animal chemistry and organization, a sufficient key for decyphering the hieroglyphics of life, health, disease and remedy. As many laborers, and as much attention given to our art as was given to Egyptian hieroglyphics, would solve the riddle of our Sphynx as well as theirs.

Shearman's optimism now seems warranted, and Bartlett's view hopelessly pessimistic. It would not be long before Pasteur would revolutionize thinking about disease in general, and pneumonia in particular, with his theory of microorganisms. One can only wonder what Bartlett's response to this theory of "ultimate" causes of disease would be. Indeed, we may still wonder if microorganisms are the *ultimate* causes of infectious diseases. Even with our recent revolutionary advances in understanding the human gene, it is still an open question whether gene function gives us the ultimate explanation of disease.

With regard to therapy, blood-letting seems to have been so well accepted that Bartlett could not question its effectiveness. It was one of the hidden assumptions that Bartlett wanted to eliminate, but could not because he was blind to it. As King (*Transformations*, 200) points out, both Bartlett and his contemporary reviewers seem to have missed the need for controls in order to substantiate claims about the efficacy of blood-letting. In fact, Bartlett did talk about the need for controls in studies of therapeutics on p. 160 of his *Essay*. In regard to blood-letting, it may well be that it did ameliorate some of the symptoms of pneumonia in some cases for some very good reasons, and that this was observed. It is highly unlikely, however, that a good case-controlled study would have substantiated this claim more generally.

The Philosophy of Therapeutics

Bartlett apparently never completed work on his *Philosophy of Therapeutics*, for the manuscript exists, literally, in fragments. The work, like his *Essay*, consists of explanations of several propositions. Bartlett had written out the propositions on one double sheet of paper and was cutting and pasting the propositions onto other sheets containing the explanations, and the process was never completed. The work began as a lecture at the College of Physicians and Surgeons, as we can judge from Bartlett's reference to his title as Professor of Materia Medica and Medical Jurisprudence (Bartlett adds "Therapeutics" to his title in the text.) Additions and deletions in Bartlett's own hand make it clear that he was transforming the lecture into an essay. For instance, "introductory lecture" is crossed out and replaced with "short essay." The pages were renumbered consecutively, and paragraph numbers were also added, obviously to make it easier to put the various paper fragments in order.

The manuscript contains twelve propositions, the tenth of which, along with its explanatory text, is completely crossed out. This edition restores the tenth proposition. It is unclear why Bartlett wanted to delete it. The last paragraph under the fifth proposition is an addition that is written in a hand other than Bartlett's. It is clearly numbered and intended to be inserted into the text at that point. This suggests that perhaps Bartlett was still working on the essay at a time when he became too ill to write for himself.

The Philosophy of Therapeutics is a consideration of "the nature, the degree of certainty and positiveness, and the sources of knowledge" of the phenomena of therapeutics. The propositions are a further elucidation of the philosophy already put forth in the *Essay*, but applied to therapeutics in particular. The first says that the science of therapeutics consists in the "phenomena analyzed and classified, which result from the curative relations existing between the materia medica, on the one hand, and the morbid actions, tendencies, and conditions of the human economy, on the other." Bartlett argues that there is no "true scientific relation" between the materia medica and therapeutics. Here, however, he admits that a knowledge of the "natural history" of the substances of the materia medica is necessary and important. It is just that a knowledge of the chemical properties of a substance does not throw light upon its therapeutic effectiveness. This skepticism goes along with Bartlett's denial that we can know ultimate causes of disease. From a twenty-first century perspective, knowledge of chemical composition and mechanisms of action of drugs seems essential for understanding therapeutic effectiveness. Then again, even in the present time we have many drugs with proven effectiveness and unknown mechanisms of action. Perhaps the most important comment to make here is that Bartlett's qualification softens somewhat the accusations that he is a rigid ideologue.

The second, third and fourth propositions have to do with the science of therapeutics as the study of the interaction of therapeutic agent and diseased state, and the degree of certainty in the science of therapeutics. Bartlett argues that the phenomena connected with therapeutic agents are uniform and hence the science of these agents is likewise uniform and relatively certain. The science of disease, however, is much less certain. Bartlett here gives a fuller account of the "numerical method," which will increase certainty in the science of disease; but he recognizes an essential individuality in cases of disease. Hence, he makes some interesting ontological claims that were not evident in his *Essay*, and sheds some new light on his previous epistemological claims. Bartlett seems to be suggesting that some uncertainty is inherent in individual cases of disease. The best that we can do is to observe as objectively as we can. Science "waits with passive and sublime indifference" for whatever the truth of the matter is.

The fifth proposition is an application of Bartlett's fundamental empirical philosophy to therapeutics. Therapeutics is a science of "pure observation." Therapeutic effectiveness cannot be deduced from pathology, or from studies on animals. The sixth proposition is a development of how such observations should be carried out, again advocating the numerical method.

The seventh and eighth propositions further explain how the inexactness of therapeutics depends on the inexactness of our knowledge of disease. The science and art of therapeutics are not "approximative" because of imperfect knowledge, but because disease is always individual and in flux. This point is not well developed, but appears again to be a softening of his previous stance on the hard uniformity of scientific laws. Bartlett continues, however, to hold our epistemological limitations as the major reason for uncertainty.

The ninth proposition is again parallel to Bartlett's claim in the *Essay*. A law or principle of therapeutics is simply a generalization of facts, phenomena, events and relations. Bartlett says that the "vital forces" upon which the phenomena depend are "beyond the reach of human knowledge," or at least "have thus far eluded all our researches and investigations." He seems to have believed that unlike laws in astronomy, we cannot advance from "lower empirical laws" to the "simple and ultimate forces from the action of which these laws have flowed, to that great, absolute higher law, which contains and included all the laws below it." The work of the life sciences and of therapeutics is, he says, the work of Kepler and not that of Newton. Bartlett seems to be unsure of what he really thinks about our epistemological powers, however. In *The Philosophy of Therapeutics* he makes his strongest statements about our epistemological limits. He says that although it is unphilosophical to attempt to limit or define the limits of our epistemological powers, those powers certainly do have limits. He ventures an opinion that the vital forces are beyond these limits. These later comments are crossed out, however, and so we can only conclude that Bartlett continued to puzzle about this difficult philosophical question to the end of his life.

The tenth proposition is crossed out in its entirety in the manuscript. It has to do with the notion of a "rational indication" for therapy, a term used by John Brown. Bartlett says that the only true indication for therapy is the "removal or mitigation of disease." It may be that Bartlett realized the obviousness of this claim, and perhaps the lack of any meaningful difference between a philosophical indication and a rational indication.

The eleventh proposition again brings up the idea of epistemological limits. The mechanism of action of our remedies, or, as he puts it, their *modus operandi*, is unknown and wholly hypothetical. Bartlett cautions against extrapolation in the use of particular agents. If blood-letting is helpful in cases of inflammation of one organ, we should not conclude that it will be helpful in cases of inflammation of other organs.

The twelfth proposition simply restates the object of the science of therapeutics: given a disease, to find a remedy.

The Philosophy of Therapeutics continues Bartlett's empiricist project, but it lacks some of the stridency and seeming dogmatism of the *Essay*, the characteristics that King finds so objectionable. Bartlett is a bit more reflective and humble about the epistemological strengths of empiricism, even venturing a bit into metaphysics to ponder epistemological limitations. In addition, this work adds to Bartlett's philosophical reflections on diagnosis and medical research his philosophical reflections on treatment. Hence, *The Philosophy of Therapeutics* ought to be seen as an essential completion of Bartlett's works on the philosophy of medicine.

BARTLETT'S OTHER WRITINGS

In his various capacities as physician, educator, and government official, Elisha Bartlett published a number of other addresses, articles, books and poems. The present listing is not encyclopedic, especially with regard to his poetry, but covers the most important of his works, especially those directly related to his philosophy of medicine and the more general philosophical positions that influenced his philosophy of medicine.

Editorial Work and Translation

Bartlett began his medical writing in the short-lived *Monthly Journal of Medical Literature and American Medical Students' Gazette*. Only three numbers of this journal were issued, in 1832, but it had a distinctive approach that was intended to supplement the more practical emphases of the existing medical journals. Beck surveys the contents of the three issues and cites Bartlett's own hopes for the journal: to "keep alive and stimulate in the young medical scholar the sometimes flagging energies of study" through presentations of "medical history, medical literature, accounts of medical institutions and hospitals, medical biography, including sketches of the character, lives, and writings of the chief masters of our art," which are excluded from the more practical journals (Beck, 126). Beck (130-33) speculates that the journal failed for many reasons, but largely because of financial difficulties and the failure of its intended readership, medical students, to sustain it. According to Huntington, Bartlett incorporated this journal, working with Dr. A. L. Pierson and Dr. J. B. Flint, into the *Medical Magazine*, which was published monthly in Boston beginning in July 1832, and continuing for the succeeding three years.

In 1831, Bartlett published a translation from the French of a small book by J. L. H. Peisse (1803-1880) entitled *Sketches of the Character and Writings of Eminent Living Surgeons and Physicians of Paris*. The book chronicles the work of nine French physicians (out of the twenty-two included in the original French edition of Peisse) and reflects the theory and practice of French medicine in the early nineteenth century, a theory and practice that would importantly influence Bartlett's own philosophy of medicine.

Bartlett also undertook some typically scholarly work when he was engaged by Horace Mann to produce a revised edition of William Paley's (1743-1805) *Natural Theology* (Patterson, 369). In the edition, he integrates notes by Lord Brougham and Sir Charles Bell, which had appeared as an appendix in a previous London edition, into the appropriate chapters of Paley's text.

Medical Writing

The History, Diagnosis, and Treatment of Typhoid and of Typhus Fever; With an Essay on the Diagnosis of Bilious Remittent and of Yellow Fever (1842) was published during Bartlett's tenure at Transylvania University. Dickson (755) considers this Bartlett's greatest work, and it later appeared in a second edition, retitled *The History, Diagnosis, and Treatment of the Fevers of the United States* (1847), a third edition in 1852, and a fourth edition, published posthumously in 1857.. The second edition expands Bartlett's treatment of periodical and yellow fevers to the extent that this material comprises one half of the volume, making the work what Bartlett (ix) calls "a Systematic and Methodical Treatise on the Fevers of the United States." Burrage (4) regards the book as "one of the most noteworthy contributions to medicine of the first half of the nineteenth century." It gives a remarkably accurate description of the physical findings and epidemiology of typhoid fever and it played an important role in enabling the differentiation of typhoid fever from typhus fever. It was not until years later that the microbial causes of these (or any) diseases were identified. Nonetheless, the book is still considered a milestone in understanding these two diseases.

In 1850, Bartlett published his *History, Diagnosis and Treatment of Edematous Laryngitis*, which was printed both in the *Western Journal of Medicine and Surgery* and as a separate pamphlet. In it, Bartlett describes the pathological lesions, varieties and forms of the disease, its natural history, epidemiology, prognosis and treatment. Bartlett is meticulous in describing the condition, much in the way that he described the fevers in his book. In a section on theory, he defends the view that the disease is inflammatory, and not just simple dropsy, by appealing to observation, which most often reveals an inflammatory component. Bartlett quotes extensively from literature published in France. One of Bartlett's contemporaries (*New York Journal of Medicine*, 83) criticizes him for maintaining the term "laryngitis" for a condition that really affects the glottis. Bartlett was, however, simply retaining the nomenclature of the condition that was prevalent in his own day. With respect to etiology, from our current perspective Bartlett's explanation of the disease seems to miss the major mark, but he can hardly be blamed for not seeing things from the theoretical viewpoint of the twenty-first century.

Bartlett also wrote in the medical journals on cases from his medical practice. One example is an 1839 detailed case report of a fifteen-year-old girl with long term severe headaches of unknown cause and for which no effective treatment could be found. Another is an 1842 account of an autopsy performed on a sixty-three-year-old woman who had died of typhoid fever. Bartlett reported the case because at the time typhoid fever was considered to be a disease of young people.

Medical Exhortation

Bartlett gave several lectures to medical audiences on various non-technical medical topics. The topics varied widely; they celebrated people and festive occasions. We might best characterize them as medical exhortations. Some of them are good expressions of parts of Bartlett's medical philosophy and, more generally, his philosophy of human nature.

In *An Address Delivered at the Anniversary Celebration of the Birth of Spurzheim and the Organization of the Boston Phrenological Society, January 1, 1838*, Bartlett discusses the progress of science of the human mind and the validity of phrenology. Phrenology was, at the time, a popular "science" of the mind, based on the theories of the Viennese physician Franz Joseph Gall (1758-1828). Phrenologists held that the brain is the organ of the mind, and that since the skull takes its shape from the brain, the surface of the skull is an accurate indicator of psychological aptitudes and character. Bartlett admits that he is not in a position to make a judgment about the truth of phrenology because he has not adequately studied it, but he puts forth a principle that would later be developed as the heart of his philosophy of medical science: that the truth of the science will ultimately be proved or disproved through observation. A second test is that any true science will ultimately be for the human good (6-7). This is a curious claim—a value judgment, and not at all an empirically verifiable claim. Bartlett praises phrenology for holding up the ideal of human equality as linked with "certain states and conditions of the several elementary principles of our spiritual being" (12). All powers of the human mind are developed only by "developing and exciting" their own particular activities. This is the role of education (14-16). Bartlett advocates, even in the public schools, religious instruction—not, to be sure, any sectarian theology, but rather the inculcation of a religious spirit and the moral dispositions of "benevolence, conscientiousness, marvellousness, hope, veneration, and ideality"—in short, a philosophy that "consists in the absolute supremacy of spiritual good over all conventional and material good." (23-24). Thus Bartlett shows that he is not a materialist and upholds his religious values; he uses the occasion to exhort his audience about social and moral ideals.

In his *Valedictory Address to the Graduating Class of Transylvania University, 1843*, Bartlett recommends several principles, which, if followed, will lead to success, respectability, usefulness and happiness in the practice of medicine. The motive of pecuniary gain, although not among the "highest and noblest," is nevertheless a "proper, a legitimate, a laudable one." Wealth is a great good that can enable one to assist the needy. Likewise, ambition, although not among the highest of motives, when "purified and ennobled by a subjection to other and loftier powers," can be laudable. Bartlett urges the graduates to cultivate a habit of "scientific and philosophical study." Every new case of disease should present the physician a new subject for investigation, composed of elements combined in new ways and under novel conditions; these can be appreciated only by a "thorough investigation, and a careful analysis, applied to that individual case." The last motive Bartlett recommends is a sense of duty, "the loftiest and noblest principle of human conduct."

A Brief Sketch of the Life, Character, and Writings of Dr. William Charles Wells, delivered before the Louisville Medical Society, December 7, 1849, is a tribute to a South Carolina physician who was politically ostracized as a Tory during the American Revolution, followed his father abroad, and took his medical degree at Edinburgh. Returning home after the war, Wells produced an essay on dew, the source of which was a controversial topic at the time. Wells went about his work scientifically, embracing an empirical view of science and medicine much like Bartlett's own. The *Literary World*, in a short notice, emphasized the humanistic side of Bartlett's address, a dimension that Bartlett had tried to bring to the forefront in his journal for medical students.

> This address was delivered with the design of showing by biography the pleasures and benefits which a practical physician might derive by cultivating tastes and studies not immediately connected with his own occupation. There is need enough for such an address, for the tendency of the medical profession is too exclusive, so much so that their abilities are judged by the fact, that while appearing to be tolerable sensible men, they are totally ignorant of everything with which the community are acquainted, and therefore, as they know nothing else, they must understand their own profession.

Bartlett's *Discourse on the Times, Character, and Writings of Hippocrates* was delivered at the College of Physicians and Surgeons, New York, at the opening of the 1852-3 term. It was subsequently published as a monograph, seventy-two pages in length. Bartlett here reflects upon Hippocrates and the character that elevated him to the pinnacle of medical practice in ancient Greece (23-27). First, Hippocrates would begin with an exposition of the most important and prevalent errors of medical doctrine and practice. Next, he would warn his hearers against the "subtle and dangerous errors of superstition." Third, he would caution his hearers about the "seductive but dangerous influences of the philosophers" and would compare them with the "humbler philosophy of observation and experience" without which there can be no advancement of real knowledge. Bartlett places the thought of Hippocrates in the context of the philosophy of the time, which was not divorced from its polytheism. This leads Bartlett (37) to conclude:

> Certainly, I need not insist upon the formidableness of the obstacles presented by this religion and philosophy, to the progress of all sound and stable medical doctrine, and of all natural science whatever; nor upon the clearness and acuteness of that insight which penetrated and dispelled the darkness of these inevitable but stupendous delusions.

Diseases were treated in the temples and in private, hygiene was studied in the gymnasia, animals were dissected, and the philosophers "endeavored to coordinate, to interpret, and to rationalize the teachings of the physicians and hygienists" (40). While Hippocrates did not literally create the science and art of medicine, he did detach it from religion, holding that diseases had natural, and not divine causes. Bartlett finds a kindred spirit in Hippocrates the empiricist. He concludes this address with a statement reflecting faith that an empirical approach to medical science will bring progress, and ultimately lead us to truth (72-73).

> The science of medicine is, historically, twenty-two centuries old. Since its origin in Greece, it has never ceased to be cultivated, wherever any considerable degree of civilization has been reached. During all this long period, the science of medicine, like its kindred sciences of observation, has obeyed its own inherent and vital law of development. Subject always to its various and complicated relations; sometimes seduced or driven from its true path; sometimes obstructed or hindered in its march; sometimes dragged backward, it has still steadily struggled onward, obedient to the living principle of growth and progress within it. . . .
>
> It is natural enough, when we look at the popular medical delusions of our day, and the skepticism as to the claims of medical science and art, which has seized upon the minds even of sensible and cultivated men,—that we should have some misgivings as to the permanency and stability of this science and art. But the great organic laws of nature are not to be suspended, nor reversed, nor turned aside. The lessons of twenty-two centuries are not to be forgotten, nor made to contradict themselves, for the first time, to-day. The science is constituted by the results of the toilsome and conscientious study of nature during those long centuries, recorded, systematized, and arranged; and as long as nature remains what it was two thousand years ago, and what it is to-day, these results will remain.

The *New York Journal of Medicine* remarked that "among the annual trashy emanations of our medical colleges, it is good occasionally to meet with a production which bears like this the impress of study and research" (92).

One can see, then, that Bartlett used his hortatory addresses in many contexts to espouse the empiricism that is at the heart of his philosophy of medicine

Civic Addresses

Bartlett also gave many lectures for non-medical audiences. Several of these were delivered in the context of his duties as mayor of Lowell or representative in the Massachusetts legislature. Very often, these addresses touched upon the topics of health and education that were the focus of his activities throughout his life.

The "Laws of Sobriety," and the "Temperance Reform:" An Address delivered before the Young Men's Temperance Society in Lowell, March 8, 1835, presents Bartlett's general philosophy of human nature. The human being is subject to certain laws concerning both physical and moral existence, and these laws reciprocally influence each other. It is a moral duty to "manage and use the body" so as to secure its "most perfect development, its freest action and its longest life" (5). The excessive or untimely use of alcohol, tea and coffee, and the "luxuries" of opium and tobacco can violate the laws of temperance and thus interfere with one's duty to maintain the body. Still, all these things can have proper uses. Distilled spirits, more than beer, wine, or even opium, are what lead to the worst of evils. Bartlett calls for a prohibition of the manufacture and use of distilled spirits. He admits that he had at one time, for several years, used tobacco, in the form of snuff and smoking, and judges that tobacco is far more injurious to health than wine. As for wine, beer (except for the "heavy, strong beer of our breweries"), coffee, tea and cider, their judicious use should be allowed as consistent with temperance, although "small beer" can cause a "flatulency" that may "sometimes prove a source of some inconvenience" (23).

In *Obedience to the Laws of Health, a Moral Duty: A Lecture Delivered before the American Physiological Society, January 30, 1838*, Bartlett again articulates his philosophy of human nature—the idea that there exists in all human nature an innate sense of moral duty. Humans have a two-fold nature: spirit and flesh. Attempting to separate these does violence to human nature. Because the body is related to the human mind, there is a duty to maintain its well-being. Although Bartlett seems to be making religious or metaphysical assertions here, he again puts his claim in the context of his empiricism. He treats the subject only insofar as it can be ascertained by observation (15). The laws of the body are only beginning to be known, but Bartlett puts great trust that these laws will be second, only to Christianity, in promoting "the truest interests of our race" (22).

This idea is repeated in *The Head and the Heart, or the Relative Importance of Intellectual and Moral Education: A Lecture delivered before the American Institute of Instruction, in Lowell, August, 1838*. Bartlett here argues that the highest element of human nature is the moral and religious element (4). All other parts of human nature should be subject to this sovereign element. He observes that although much attention is given to religious practice, insufficient stress is placed upon moral education. The way to overcome moral evil is not though intellectual education, however, but though education of the heart. "Not philosophy, but conscience,— not science, but religion, is the minister and physician to the mind so diseased" (14).

In *A Lecture on the Sense of the Beautiful, Delivered before the Lexington Lyceum, January 20, 1843*, Bartlett argues that human beings have always recognized an innate sense of complete, faultless perfection. Although this faculty is not the highest and noblest power, it is still among the highest and noblest. Bartlett's argument is simply an appeal to all people to attend to their own experience to confirm this claim. Beauty is always attractive; ugliness is always repulsive. This faculty can be educated and strengthened only by placing oneself in relation to beautiful things. Bartlett eloquently describes the beauty of earth, sky, seas and rivers, and animal life, with human life at its crown. Understanding the laws of nature, which govern all these things, is the task of science, and such understanding naturally produces in us a recognition of beauty. Poetry, likewise, produces such feelings in us. The beauty of the world as well as the human capacity to appreciate it is all a creation of God.

As might be expected, Bartlett, as a mayor and legislator, was deeply interested in the social issues of his day, even those not directly touching on health care. *An Oration Delivered to the Municipal Authorities and the Citizens of Lowell, July 4, 1848*, is a discussion of the state of the nation, its people and its relations with the countries of Europe. Bartlett condemns slavery and speaks of three great lessons that the United States has taught the world: the human capability for self-government; the absence of the necessity, in a free state, for a standing army; and the claim that religion, the highest of human interests, is most effectively promoted by its complete separation from the state.

Perhaps Bartlett's most well-known pamphlet, *A Vindication of the Character and Condition of the Females Employed in the Lowell Mills Against the Charges Contained in the Boston Times and the Boston Quarterly Review* was originally published in July 1839 in the *Lowell Courier*. The burgeoning mill towns of the early nineteenth century required cheap housing for the workers who poured into them from the surrounding countryside. Many worried that the female boarding houses would lead to, as Cassedy (*Medicine and American Growth*, 165) puts it, "seductions and licentiousness." To answer such charges, which were being made in the Boston press against the boarding houses of Lowell, Bartlett used the results of questionnaires and statistical analysis to defend the moral character and the health of the young women employed in the Lowell mills.

Poetry

Elisha Bartlett walked in literary circles as well as medical ones. Besides his friendship with Oliver Wendell Holmes, Bartlett had at least a passing acquaintance with other prominent authors. Bartlett had met Charles Dickens in 1842, while he was teaching at Transylvania, and had sent both Dickens and Washington Irving copies of some of his works. Winterich, in his short book of 1933, documents Bartlett's correspondence with Dickens and Irving, and includes facsimiles of the thank-you notes sent by both to Bartlett.

This interest in poetry was not just a diversion for Bartlett. Rather, Bartlett championed the humanities as essential for the development of a competent and compassionate physician. In his introduction to the first issue of his *Monthly Journal of Medical Literature*, Bartlett put forth his intent to help medical students "enlarge and liberalize" their minds. "If our profession ever vindicates its legitimate claim to the appellation of *liberal*, it must be cultivated with some other than the single purpose of getting for services rendered, an equivalent in fees."

Bartlett began publishing his poetry in the periodicals of his day—newspapers such as the *Massachusetts Spy* and the *Yeoman's Gazette*—identified only as "E." We know that these poems are his from a letter that Bartlett wrote to his sister in 1828 (Y1:3:1) listing eighteen published poems and speaking of others that he had forgotten. He had begun publishing his poetry as early as 1822 and continued publishing poetry during the course of his medical studies. He wrote of storms at sea and piracy, ships of sail and steam, a mother's joy and the passage of time. Even at the age of eighteen, Bartlett seemed to sense the seriousness of the suffering he would encounter in his medical practice, and perhaps even in his own illness. He wrote in an untitled poem in the November 13, 1822, *Massachusetts Spy*:

How many Bards have sung of youth,
When all was novelty and truth;
And deck'd in garb of fairy dress,
Its days of light and loveliness.

Reverse the picture—time will show
Its flitting shades of joy and wo;
And though its smiles are bright to-day,
To-morrow sees them fade away.

Bartlett continued to write poetry until the end of his life. He worked on *Simple Settings, in Verse, for Six Portraits and Pictures from Mr. Dickens's Gallery* (1855) while he was in the grip of his terminal illness. Bartlett wrote in a short preface to that book:

> Dear Friend,
> I send you a copy of a few verses which I have had printed and put into covers, as a Christmas gift. The inditing of them has been to me a most pleasant occupation,—I cannot call it a labor,—and has helped me to while away and fill many an hour, that would otherwise have been weary or vacant, in my invalid life.

Much of the verse in that collection seems to be a reflection on the passing of life and the hope of a heavenly reward. Oliver Wendell Holmes, in his obituary of Bartlett (52), writes of this work:

> When to the friends he had loved, there came as a farewell gift not a last effort of the learning and wisdom they had been taught to expect from him, but a little book with a few songs in it, songs with his whole warm heart in them, they knew that his hour was come, and their tears fell fast as they read the loving thoughts that he had clothed in words of natural beauty and melody. The cluster of evening primroses had opened, and the night was close at hand.

Elisha Bartlett never forgot that medicine is more than the empirical science that he so ardently advocated. His philosophy of medicine is rooted in a profoundly humanistic and even religious philosophy of the person. A few lines of verse from "Poor Jo" (*Simple Settings*, 43) serve as fitting conclusion to this introduction to Bartlett's philosophy of medicine:

> Unveil the mysteries that lie
> On earth, in ocean, air, and sky,
> And open every ear and eye,
>
> To see the beauty that unsprings,
> To hear the music that outrings,
> From all this wondrous frame of things,
>
> O'er all His works, with glory lit,
> To know the meaning, God has writ
> In characters so fair and fit.

BIBLIOGRAPHY

PRIMARY SOURCES

Published Works of Elisha Bartlett

The following is a chronological listing of Bartlett's most important publications.

Bartlett, Elisha. *"Laws of Sobriety", and the "Temperance Reform": An Address Delivered before the Young Men's Temperance Society in Lowell, March 8, 1835*. Lowell: Published by the Society, 1835.

———. *An Address Delivered at the Anniversary Celebration of the Birth of Spurzheim, and the Organization of the Boston Phrenological Society, January 1st, 1838*. Boston: Marsh, Capen and Lyon, 1838.

———. *The Head and the Heart, or the Relative Importance of Intellectual and Moral Education: A Lecture Delivered before the American Institute of Instruction in Lowell, August, 1838*. Boston: William D. Ticknor, 1838. Reprint, Philadelphia: Anti-Slavery Office, 1844.

———. *Obedience to the Laws of Health, a Moral Duty: A Lecture Delivered before the American Physiological Society, January 30, 1838*. Boston: Julius A. Noble, 1838.

———. "Chronic Cerebral Affection: Long Intense Headaches: Double Consciousness: Extraordinary Memory of Events: Inefficacy of Treatment: Diagnosis Doubtful." *American Journal of the Medical Sciences* 24, no. 47 (1839): 49-58.

———. *An Introductory Lecture on the Objects and Nature of Medical Science, Delivered in the Hall of the Medical Department at Transylvania University, November 3, 1841*. Lexington, Ky.: N. L. and J. W. Finnell, Printers, 1841.

———. *Vindication of the Character and Condition of the Females Employed in the Lowell Mills: Against the Charges Contained in the Boston Times, and the Boston Quarterly Review*. Lowell: Leonard Huntress, Printer, 1841.

———. *The History, Diagnosis, and Treatment of Typhoid and of Typhus Fever; With an Essay on the Diagnosis of Bilious Remittent and of Yellow Fever*. Philadelphia: Lea and Blanchard, 1842.

———. "Typhoid Fever in a Patient Sixty-Three Years Old." *Boston Medical and Surgical Journal* 27 (1842): 82-84. Reprint, *Medical Examiner* 1, n.s. (1842): 620-622.

———. *A Lecture on the Sense of the Beautiful, Delivered before the Lexington Lyceum, January 20, 1843*. Lexington, Ky.: James Virden, Book and Job Printer, 1843.

———. *Valedictory Address to the Graduating Class of Transylvania University, 1843*. Lexington, Ky.: Lexington Intelligencer Print, [1843?]. A section reprinted as "Moral Duties of the Physician." *Boston Medical and Surgical Journal* 28 (1843): 420-421.

———. *An Essay on the Philosophy of Medical Science*. Philadelphia: Lea and Blanchard, 1844.

———. "Letter from Paris, April 20th, 1846." *Western Lancet* 5 (1846): 107-111.

———. "Letter from Paris, May 4th, 1846." *Western Lancet* 5 (1846): 172-176.

———. *History, Diagnosis, and Treatment of the Fevers of the United States*. Philadelphia: Lea and Blanchard, 1847. 2d ed., revised, of *The History, Diagnosis, and Treatment of Typhoid and of Typhus Fever; With an Essay on the Diagnosis of Bilious Remittent and of Yellow Fever.*

———. *An Inquiry into the Degree of Certainty of Medicine, and into the Nature and Extent of Its Power over Disease*. Philadelphia: Lea and Blanchard, 1848.

———. *An Oration Delivered to the Municipal Authorities and the Citizens of Lowell, July 4, 1848.* Lowell: Committee of Arrangements, 1848.
———. *A Brief Sketch of the Life, Character, and Writings of Dr. William Charles Wells, Delivered before the Louisville Medical Society, December 7, 1849.* Philadelphia: Lea and Blanchard, 1849.
———. "The History, Diagnosis and Treatment of Edematous Laryngitis." *Western Journal of Medicine and Surgery* 5, 3d ser. (1850): 209-240.
———. "On the Causes of Edematous Laryngitis." *New York Journal of Medicine* 4, n.s. (1850): 398-399. Excerpt from the *Western Journal of Medicine and Surgery.*
———. "On the Diagnosis of Edematous Laryngitis." *New York Journal of Medicine* 4, n.s. (1850): 403-404. Excerpt from the *Western Journal of Medicine and Surgery.*
———. *Discourse on the Times, Character, and Writings of Hippocrates: Read before the Trustees, Faculty and Medical Class of the College of Physicians and Surgeons, at the Opening of the Term of 1852-3.* New York: H. Bailliere, 1852.
———. *History, Diagnosis, and Treatment of the Fevers of the United States.* 3d ed. Philadelphia: Blanchard and Lea, 1852.
———. *Simple Settings, in Verse, for Six Portraits and Pictures from Mr. Dickens's Gallery.* Boston: Ticknor and Fields, 1855.
———. *History, Diagnosis, and Treatment of the Fevers of the United States.* 4th ed. Revised by A. Clark, M.D. Philadelphia: Blanchard and Lea, 1856.

Works Translated or Edited by Elisha Bartlett

Peisse, J. L. H. P. *Sketches of the Character and Writings of Eminent Living Surgeons and Physicians of Paris.* Translated by Elisha Bartlett. Boston: Carter, Hendee and Babcock, 1831.
Monthly Journal of Medical Literature, January-March 1832.
Medical Magazine (edited with A. L. Pierson and J. B. Flint) July 1832-July 1835.
William Paley. *Paley's Natural Theology: With Selections from the Illustrative Notes, and the Supplementary Dissertations, of Sir Charles Bell, and Lord Brougham. The Whole Newly Arranged, and Edited by Elisha Bartlett, M.D.* 2 vols. Boston: Marsh, Capen, Lyon, and Webb, 1839.

Selected Unpublished Manuscripts

Manuscripts in this bibliography are listed alphabetically by their library source according to the following library abbreviations.

B = Brown University Library. John Hay Library. Published by permission of Brown University Library.
H = Harvard University Libraries. Houghton Library. Published by permission of the Houghton Library, Harvard University.
R = University of Rochester Library. Rush Rhees Library. Published by permission of the Department of Rare Books and Special Collections, University of Rochester Library.
Y = Yale University Library. Sterling Library. Published by permission of Manuscripts and Archives, Yale University Library.

The numbering system for the references used in this volume differs a bit according to each library. For Brown University, the first two numbers refer to the Brown Library's manuscript call number of the indicated collection. The third is simply a number assigned consecutively to the manuscripts chosen from the indicated collection.

The Harvard University Library gives each manuscript within a collection its own numbered folder. For the Oliver Wendell Holmes papers (call number bMs Am 1241.1), the letters chosen for inclusion are numbered from 100 to 109. They are referred to in the text of this volume as H100, H101, etc. The letter to Augustus A. Gould (bMs Am 1210) is referred to as H1.

The major collections of Elisha Bartlett's papers are owned by the University of Rochester and Yale University. These two collections contain a wealth of information in Bartlett's correspondence, lecture notes, manuscripts and poetry. Material listed in this bibliography is selected as especially relevant to Bartlett's philosophy of medicine, medical education and teaching. The following system is used for referring to the documents from Yale University and the University of Rochester. After the letter designating the library, three numbers are given. The first refers to the box number of the collection and the second to the folder number. The manuscripts selected for inclusion are then numbered consecutively; this is the third number of the citation. For example, the third letter selected from box 1, folder 12 at Yale would be Y1:12:3.

Brown University Library

Oliver Wendell Holmes Papers (Ms.15.129)

B15:129:1 Oliver Wendell Holmes to Elisha Bartlett, Boston, Massachusetts, 21 November 1838.

B15:129:2 Oliver Wendell Holmes to Elisha Bartlett, Hanover, New Hampshire, 26 September 1839.

Edward Everett Papers (Ms.51.43)

B51:43:1 Edward Everett to the Trustees of the College of Physicians and Surgeons of the State of New York, Boston, Massachusetts, 7 March 1840.

Harvard University Libraries

Oliver Wendell Holmes Papers (bMs Am 1241.1 (100-109))

H100 Elisha Bartlett to Oliver Wendell Holmes, Lowell, Massachusetts, 3 November 1838.

H101 Elisha Bartlett to Oliver Wendell Holmes, Lowell, Massachusetts, 10 April 1839.

H102 Elisha Bartlett to Oliver Wendell Holmes, Pittsfield, Massachusetts, 27 October 1839

H103 Elisha Bartlett to Oliver Wendell Holmes, Lowell, Massachusetts, 12 June 1842.

H104 Elisha Bartlett to Oliver Wendell Holmes, Lexington, Kentucky, 6 February 1844.

H105 Elisha Bartlett to Oliver Wendell Holmes, Baltimore, Maryland, 4 November 1844.

H106 Elisha Bartlett to Oliver Wendell Holmes, Slatersville, Rhode Island 6 May 1845.

H107 Elisha Bartlett to Oliver Wendell Holmes, Lexington, Kentucky, 3 December 1846.

H108 Elisha Bartlett to Oliver Wendell Holmes, Woonsocket, Rhode Island, 9 September 1848.

H109 Elisha Bartlett to Oliver Wendell Holmes, Woonsocket, Rhode Island, 28 June 1849.

Augustus A. Gould Papers (bMs Am 1210)

H1 Elisha Bartlett to Augustus Addison Gould, Woodstock, Vermont, 22 May 1850.

University of Rochester Library

Bartlett Family Papers (D.96)

R1:1:1 Elisha Bartlett to Charles Dickens, n.p., n.d.

R1:2:1 Elisha Bartlett to Caroline Bartlett, Worcester, Massachusetts, 16 November 1824.

R1:2:2 Elisha Bartlett to Caroline Bartlett, Boston, Massachusetts, 20 November 1825.

R1:3:1 Elisha Bartlett to John Orne Green, M.D., Lowell, Massachusetts, 25 November 1833.

R1:4:1 Elisha Bartlett to John O[rne] Green, Pittsfield, Massachusetts, 1 November 1835.

R1:4:2 Elisha Bartlett to John Orne Green, Pittsfield, Massachusetts, 2 October 1836.
R1:5:1 Elisha Bartlett to John Orne Green, Hanover, New Hampshire, 8 September 1839.
R1:5:2 Elisha Bartlett to John Orne Green, Lexington, Kentucky, 21 December 1841.
R1:6:1 Elisha Bartlett to John O[rne] Green, Lexington, Kentucky, 31 March 1843.
R1:6:2 William E. A. Aikins to Elisha Bartlett, Baltimore, Maryland, 8 April 1844.
R1:6:3 N[athan] R[yno] Smith to Elisha Bartlett, Baltimore, Maryland, 16 April 1844.
R1:6:4 G[ilman] Kimball to Elisha Bartlett, Lowell, Massachusetts, 19 November 1844.
R1:7:1 Elisha Bartlett to Oliver Wendell Holmes, Baltimore, Maryland, 13 January 1845.
R1:7:2 Roby to Elisha Bartlett, Boston, Massachusetts, 15 March 1845.
R1:7:3 Robert Peter to Elisha Bartlett, Lexington, Kentucky, 15 September 1845.
R1:7:4 B[enjamin] W[inslow] Dudley to Elisha Bartlett, Lexington, Kentucky, 30 September 1845.
R1:8:1 C. Butterfield to Elisha Bartlett, Columbus, Ohio, 23 January 1846.
R1:8:2 Elisha Bartlett to John O[rne] Green, London, 17 June 1846.
R1:8:3 Robert Peter to Elisha Bartlett, Lexington, Kentucky, 24 August 1846.
R1:8:4 Elisha Bartlett to L[eonidas] M[erion] Lawson, Lexington, Kentucky, 8 January 1847.
R1:8:5 Elisha Bartlett to John O[rne] Green, Lexington, Kentucky, 18 March 1847.
R1:8:6 John W[illiam] Draper to Elisha Bartlett, New York, New York, 3 July 1847.
R1:10:1 Peter Smith to Elisha Bartlett, Nashville, Tennessee, 25 January 1848.
R1:10:2 Alonzo Child to Elisha Bartlett, St. Louis, Missouri, 22 February 1848.
R1:10:3 J. Cobb to Elisha Bartlett, Louisville, Kentucky, 13 March 1849.
R1:11:1 John O[rne] Green to Elisha Bartlett, Lowell, Massachusetts, 2 April 1850.
R1:11:2 John William Draper to Elisha Bartlett, New York, New York, 30 July 1850.
R1:12:1 John William Draper to Elisha Bartlett, New York, New York, 12 August 1850.
R1:12:2 Benjamin Silliman, Jr., to Elisha Bartlett, New Haven, Connecticut, 14 August 1850.
R1:12:3 W[illiam] B[rown] Maclay to Elisha Bartlett, New York, New York, 19 September 1850.
R1:12:4 Tom L. Winston to Elisha Bartlett, Harlesville, Tennessee, 9 October 1850.
R1:13:1 Charles Dickens to Elisha Bartlett, London, 26 December 1850.
R1:14:1 Roby to Elisha Bartlett, West Newton, Massachusetts, 4 September 1851.
R1:17:1 B. R. Palmer to Elisha Bartlett, Woodstock, Vermont, 11 April 1853.
R1:17:2 C. R. Gilman to Elisha Bartlett, New York, New York, 13 July 1853.
R1:18:1 George Bartlett to Elisha Huntington, 17 December 1855.
R2:1:1 Address to Classmates and Fellow Students, September 1821.
R2:8:1 Philosophy of Therapeutics, undated manuscript fragment.

Yale University Library

Elisha Bartlett Papers (MS 1279)

Y1:1:1 Elisha Bartlett to Caroline Bartlett, New York, New York, 14 June 1826.
Y1:1:2 Elisha Bartlett to Caroline Bartlett, Paris, 27 September 1826.
Y1:1:3 Elisha Bartlett to Caroline Bartlett, Leghorn, Italy, 28 December 1826.
Y1:2:1 Elisha Bartlett to Otis Bartlett, Paris, 17 April 1827.
Y1:2:2 Elisha Bartlett to Caroline Bartlett, Liverpool, 8 June 1827.
Y1:2:3 Elisha Bartlett to Caroline Bartlett, Lowell, Massachusetts, 25 December 1827.
Y1:3:1 Elisha Bartlett to Caroline Bartlett, Lowell, Massachusetts, 31 May 1828.
Y1:4:1 Elisha Bartlett to "My Dear Friend," Pittsfield, Massachusetts, 9 September 1838.
Y1:4:2 Horace Mann to the College of Physicians and Surgeons, Boston, Massachusetts, 9 March 1840.
Y1:4:3 B[enjamin] Silliman to the College of Physicians and Surgeons, New Haven, Connecticut, 10 March 1840.
Y1:5:1 Elisha Bartlett to Robert Peter, Lowell, Massachusetts, 6 April 1841.
Y1:7:1 N. R. Smith to Elisha Bartlett, Baltimore, Maryland, 26 February 1843.
Elisha Bartlett to Smith, Lexington, Kentucky, 10 March 1843 (on reverse).
Y1:8:1 N. R. Smith to Elisha Bartlett, Baltimore, Maryland, 28 February 1844.
Y1: 9:1 Elisha Bartlett to John O[rne] Green, Paris, 12 July 1845.
Y1:10:1 Elisha Bartlett to George Bartlett, Rome, 19 January 1846.
Y1:10:2 Elisha Bartlett to George Bartlett, Rome, 14 February 1846.
Y1:10:3 Robert Peter to Elisha Bartlett, Lexington, Kentucky, 2 February 1846.
Y1:10:4 Robert Peter to Elisha Bartlett, Lexington, Kentucky, 2 March 1846.

Y1:10:5 Elisha Bartlett to George Bartlett, London, 31 May 1846.
Y1:11:1 J. Cobb to Elisha Bartlett, Louisville, Kentucky, 2 January 1848.
Y1:11:2 T. S. Bell to Elisha Bartlett, Louisville, Kentucky, 7 February 1848.
Y1:12:1 J. Cobb to Elisha Bartlett, Louisville, Kentucky, 2 March 1849.
Y1:12:2 W. S. Vernon to Elisha Bartlett, Louisville, Kentucky, 13 March 1849.
Y1:13:1 T. ?. Beck to Elisha Bartlett, Albany, New York, 5 March 1852.
Y2:15:1 Elisha Bartlett's notes from Paris medical lectures, December 1827-February 1828.
Y2:16:1 Elisha Bartlett's notes from James Jackson's medical lectures, 1824-5.
Y2:17:1 Elisha Bartlett's notes from Dr. Bigelow's medical lectures, 1824-5.
Y2:18:1 Manuscript fragments, undated.

SECONDARY SOURCES

Biographies of Elisha Bartlett

Arnold, James N. "Smithfield Friends Records." In *Rhode Island Friends Records*. Vol. 7. Providence, 1895?

"Bartlett, Elisha." In *The National Cyclopaedia of American Biography, Being the History of the United States*. Vol. 12, 70. New York: James T. White & Co., 1904.

Bartlett, Thomas Edward. *The Bartletts: Ancestral, Genealogical, Biographical, Historical: Comprising an Account of the American Progenitors of the Bartlett Family with Special Reference to the Descendants of John Bartlett of Weymouth and Cumberland.* New Haven: Press of the Stafford Print Co., 1892.

Burrage, Walter Lincoln. "Bartlett, Elisha." In *Dictionary of American Biography.* Vol. 2, edited by Allen Johnson, 3-5. New York: Charles Scribner's Sons, 1929.

"Death of Dr. Elisha Bartlett." *Boston Medical and Surgical Journal* 52 (1855): 507.

"Death of Elisha Bartlett, M.D." *New York Journal of Medicine* 15, n.s. (1855): 442.

Dickson, Samuel Henry. "Elisha Bartlett." In *Lives of Eminent American Physicians and Surgeons of the Nineteenth Century*, edited by Samuel D. Gross, 732-756. Philadelphia: Lindsay & Blakiston, 1861.

Holmes, Oliver Wendell. "The Late Dr. Elisha Bartlett." *Boston Medical and Surgical Journal* 53 (1855): 49-52.

Huntington, Elisha. *An Address on the Life, Character and Writings of Elisha Bartlett, M.D., M.M.S.S. before the Middlesex North District Medical Society, December 26, 1855.* Lowell: S. J. Varney, 1856.

Miller, Keith L. "Bartlett, Elisha." In *American National Biography*, edited by John A. Garraty and Mark C. Carnes, 275-276. New York: Oxford University Press, 1999.

Osler, William. "A Rhode Island Philosopher: An Address Delivered before the Rhode Island Medical Society, December 7, 1899." In *An Alabama Student and Other Biographical Essays*. New York: Oxford University Press, American Branch, 1908, 108-158. Originally published in *Transactions of the Rhode Island Medical Society* 6 (1899): 15-46. Reprinted in *Rhode Island Medical Journal* 70 (1987): 449-463.

———. "Bartlett, Elisha." In *Dictionary of American Medical Biography*, edited by Howard A. Kelly and Walter L. Burrage, 65-66. Boston: Milford House, 1928. Reprint, 1970.

Patterson, D. N. "Reminiscences of the Early Physicians of Lowell and Vicinity." In *Contributions of the Old Residents' Historical Association, Lowell, Mass.* Vol. 2, no. 4, 361-371. Lowell, The Association, 1883.

Rider, Sidney, S. *A Brief Memoir of Dr. Elisha Bartlett with Selections from his Writings and a Bibliography of the Same*. Privately printed, an edition of 300 copies. Providence: Sidney S. Rider, 1878.

Steere, Thomas. *History of the Town of Smithfield: From Its Organization in 1730-1 to Its Division in 1871*. Providence: E. L. Freeman & Co., 1881.

Reviews and Notices of Bartlett's Works

The following list of reviews and notices is chronological by date of publication of Bartlett's work. Under the title of Bartlett's work, reviews are listed alphabetically by author, when known, or by journal title. Many of the reviews in the nineteenth century medical literature are unsigned. Sometimes only the initials of the author are given. Entries give the names of authors in brackets when they can be ascertained.

The Head and the Heart, or the Relative Importance of Intellectual and Moral Education: A Lecture Delivered before the American Institute of Instruction in Lowell, August, 1838.

- *American Quarterly Register* 12 (1839): 186-187.

An Introductory Lecture on the Objects and Nature of Medical Science, Delivered in the Hall of the Medical Department at Transylvania University, November 3, 1841.

- *Boston Medical and Surgical Journal* 25 (1842): 387-389.

Vindication of the Character and Condition of the Females Employed in the Lowell Mills: Against the Charges Contained in the Boston Times, and the Boston Quarterly Review. (1841)

- *Christian Examiner* 31 (1841): 136.

The History, Diagnosis, and Treatment of Typhoid and Typhus Fever; with an Essay on the Diagnosis of Bilious Remittent and of Yellow Fever. (1842)

- *Boston Medical and Surgical Journal* 27 (1842): 171-172.
- *Medical Examiner and Record of Medical Science* 1, n.s. (1842): 693-695.

An Essay on the Philosophy of Medical Science. (1844)

- Allan, James S. "A Review of *An Essay on the Philosophy of Medical Science*, by Elisha Bartlett, M.D." *Southern Literary Messenger* 11 (1845): 330-340.
- Annan, Samuel. "*Philosophy of Medical Science.*" *Southern Literary Messenger* 11 (1845): 609-616.
- *Buffalo Medical Journal and Monthly Review* 2 (1846): 11-22.
- *Boston Medical and Surgical Journal* 31 (1844): 164.
- *Boston Medical and Surgical Journal* 32 (1845): 466.
- *Boston Medical and Surgical Journal* 33 (1845): 103, 262-263.
- *British Foreign and Medical Review* 20, no. 39 (1845): 140-147.
- L[awson, Leonidas Moreau]. *Western Lancet* 3 (1844): 386-388.
- L[awson, Leonidas Moreau]. *Western Lancet* 3 (1844): 193-198. (Letter from Elisha Bartlett and response by L[awson].)
- L[ee], C[harles] A. *New York Journal of Medicine and the Collateral Sciences* 4 (1845): 65-82.
- Leigh, Edwin. *The Philosophy of Medical Science: Considered with Special Reference to Dr. Elisha Bartlett's "Essay on the Philosophy of Medical Science." Boston Medical and Surgical Journal* 48 (1853): 69-74, 89-95, 115-121. Reprint, Boston: Ticknor, Reed, & Fields, 1853.
- *Medical Examiner and Record of Medical Science* 7 (1844): 246-247.
- N[ott], J[osiah] C. *New Orleans Medical and Surgical Journal* 1 (1844): 490-494.

The History, Diagnosis, and Treatment of the Fevers of the United States. (1847)

- *British and Foreign Medical Review* (1857): 126-127.
- *Charleston Medical Journal and Review* 3 (1848): 170-185.
- H. *Western Lancet* 7 (1848): 33
- *New York Journal of Medicine* 2, 3d ser., (1857): 126-127.

An Inquiry into the Degree of Certainty in Medicine; and into the Nature and Extent of its Powers over Disease. (1848)

- *Boston Medical and Surgical Journal* 39 (1848): 183-184.
- L. *Western Lancet* 9 (1849): 41-42.
- *Medical Examiner and Record of Medical Science* 4, n.s. (1848): 686-689.
- Shearman, J. H. *New York Journal of Medicine* 2 (1849): 80-90.
- Stillé, Albert. *American Journal of the Medical Sciences* 16, n.s. (1848): 398-406.

A Discourse on the Life and Labors of Dr. H. Charles Wells, the Discoverer of the Philosophy of Dew, Which Was Delivered before the Louisville Medical Society, December 7, 1849.
•*Literary World* 6 (1850): 127.
•*Western Journal of Medicine and Surgery* 5, 3d ser. (1850): 91-92.
History, Diagnosis, and Treatment of Edematous Laryngitis. (1850)
•*New York Journal of Medicine* 5, n.s. (1850): 83-86.
A Discourse on the Times, Character and Writings of Hippocrates, read before the Trustees, Faculty, and Medical Class of the College of Physicians and Surgeons, at the Opening of the Term, 1852-3.
•*Medical Examiner and Record of Medical Science* 9, n.s. (1853): 166.
•*New York Journal of Medicine* 10, n.s. (1853): 92-95.

Other Secondary Sources

Ackerknecht, Erwin H. "Elisha Bartlett and the Philosophy of the Paris Clinical School." *Bulletin of the History of Medicine* 24 (1950): 43-60.
———. *Medicine at the Paris Hospital.* Baltimore: Johns Hopkins University Press, 1967.
Bauer, Donald deF. "Elisha Bartlett, a Distinguished Physician with Complete Transposition of the Viscera." *Bulletin of the History of Medicine* 17 (1945): 85-92.
Bean, William B. "Elisha Bartlett—His Views of Benjamin Rush." *Archives of Internal Medicine* 116 (1965): 321-325.
Beck, Irving A. "An Early American Journal Keyed to Medical Students: A Pioneer Contribution of Elisha Bartlett." *Bulletin of the History of Medicine* 40 (1966): 124-134.
Bonner, Thomas Neville. *Becoming a Physician: Medical Education in Great Britain, France, and the United States, 1750-1945.* New York: Oxford University Press, 1995.
Bynum, W. F. *Science and the Practice of Medicine in the Nineteenth Century.* Cambridge: Cambridge University Press, 1994.
Cassedy, James H. *American Medicine and Statistical Thinking, 1800-1860.* Cambridge: Harvard University Press, 1984.
———. *Medicine and American Growth, 1800-1860.* Madison: University of Wisconsin Press, 1986.
Coburn, Frederick W. *History of Lowell and Its People.* New York: Lewis Historical Publishing Co., 1920.
Cowley, Charles. *Illustrated History of Lowell.* Boston: Lee and Shepard, 1868.
Daniels, George H., Jr. "Finalism and Positivism in Nineteenth Century American Physiological Thought." *Bulletin of the History of Medicine* 38 (1964): 343-363.
Foucault, Michel. *The Birth of the Clinic: An Archaeology of Medical Perception.* Translated by A. M. Sheridan Smith. New York: Vintage Books, 1975.
Holmes, Oliver Wendell. "Some of My Early Teachers." In *Medical Essays, 1842-1882*, 420-440. Boston: Houghton, Mifflin and Co., 1892.
King, Lester S. "Medical Philosophy, 1836-1844. In *Medicine, Science and Culture: Historical Essays in Honor of Owsei Temkin*, edited by Lloyd G. Stevenson and Robert P. Multhauf, 143-159. Baltimore: Johns Hopkins Press, 1968.
———. *Medical Thinking: A Historical Preface.* Princeton: Princeton University Press, 1982.
———. *Transformations in American Medicine: From Benjamin Rush to William Osler.* Baltimore: Johns Hopkins University Press, 1991.
Osler, William. "The Influence of Louis on American Medicine." In *An Alabama Student and Other Biographical Essays*, 189-210. New York: Oxford University Press, American Branch, 1908.
Rosenberg, Charles E. *Explaining Epidemics and Other Studies in the History of Medicine.* Cambridge: Cambridge University Press, 1992.
Sozinskey, Thomas S. "A Case of Transposition of the Heart and Other Organs, Accompanied with Bronchial Catarrh." *Medical and Surgical Reporter* 48 (1883): 1-2.
Stempsey, William E. *Disease and Diagnosis: Value-Dependent Realism.* Dordrecht: Kluwer Academic Publishers, 1999.
Starr, Paul. *The Social Transformation of American Medicine.* New York: Basic Books, 1982.
Waite, Frederick Clayton. *The Story of a Country Medical College.* Montpelier, Vermont Historical Society, 1945.

Warner, John Harley. *The Therapeutic Perspective: Medical Practice, Knowledge, and Identity in America, 1820-1885*. Princeton, N.J.: Princeton University Press, 1997.

———. *Against the Spirit of System: the French Impulse in Nineteenth-Century American Medicine*. Princeton, N.J.: Princeton University Press, 1998.

Winterich, John T. *An American Friend of Dickens*. New York: Thomas F. Madigan, 1933.

AN ESSAY ON THE PHILOSOPHY OF MEDICAL SCIENCE

"I TRUST THAT I HAVE GOT HOLD OF MY PITCHER BY THE RIGHT HANDLE."
John Joachim Beccher, Whewell's History of the Inductive Sciences, vol. iii, p. 121.

BY

ELISHA BARTLETT, M.D.

PROFESSOR OF THE THEORY AND PRACTICE OF MEDICINE, IN THE UNIVERSITY OF MARYLAND

PHILADELPHIA:

LEA & BLANCHARD

1844

Entered according to Act of Congress, in the year 1844,

By ELISHA BARTLETT,

in the Clerk's Office of the District Court of the District of Massachusetts

BOSTON:

PRINTED BY FREEMAN AND HOLLES,

WASHINGTON STREET

TO

P. CH. A. LOUIS,[1]

PHYSICIAN TO THE HOTEL DIEU; PRESIDENT OF THE SOCIETY OF MEDICAL OBSERVATION OF PARIS; AUTHOR OF RESEARCHES ON PHTHISIS, TYPHOID FEVER, PULMONARY EMPHYSEMA, ETC. ETC.

ALLOW me, my dear sir, in this public manner, to return you my warmest thanks for the readiness and the kindness, with which you assented to my request, that I might be allowed the pleasure and the privilege of dedicating this essay to yourself: — and that it will be found not altogether unworthy of such distinction, I may venture to hope, for this reason, if for no other, that it endeavors to illustrate, to develop, and to vindicate those principles of medical philosophy, which lie at the foundation of your own various and invaluable researches, — researches, the institution and publication of which have constituted a new and great era in the history of medical science.

With feelings of the highest regard,

I am, very sincerely, your friend,

E. BARTLETT.

SEPTEMBER 1, 1844.

PREFACE

I COMPLY with the custom of writing a formal preface, only for the purpose of making one or two remarks, which may be more properly made here, than anywhere else, — referring particularly to the title and the subject of my book. This title is not new; but it is the only one at all suitable for the work to which it is applied, and I had no alternative but to adopt it. I wish to say further, that my essay has no resemblance, whatever, either in design or execution to the *Essay on Medical Philosophy*, published a few years ago, by M. Bouillaud;[2] and that it differs, not less widely, from all the formal treatises, that I have been able to obtain, upon the subjects with which it is concerned. The *Elements of Medical Logic*, by Sir Gilbert Blane,[3] need no commendation from me; they are admirable as far as they go, but they embrace only a small segment of the entire circle [v] of medical philosophy. Dr. William Hillary[4] published, in 1761, "*An Inquiry into the means of improving Medical Knowledge, by examining all those methods, which have hindered or increased its improvement, in all past ages;*" a book which is strongly marked by many of the faults, which it is one of the principal objects of this essay to exhibit; mixed up, however, with much that is excellent and true. I regret, especially, not having been able to procure the two works with the following titles; — "*Traité de philosophie médicale, ou Exposition des vérités générales et fondamentales de la medecine ;*" by T. Auber;[5] Paris, 1839; and "*Novum Organum Medicorum ; — a new Medical Logic, or the art of thinking and right reasoning applied to practical medicine,*" etc. By Vicenzo Lanza, M. D.[6] of Naples. The title is all that I have seen of the first; there is a short notice of an English translation of the second, by C. Stormont, M. D. in Vol. X. of the *London Lancet*, from which I am led to believe that its fundamental doctrines are sound and philosophical. [vi]

CONTENTS

PART I.
PHILOSOPHY OF PHYSICAL SCIENCE

PRIMARY PROPOSITIONS, ... 3, 4

CHAPTER I.

Object of Essay. All science consists exclusively in phenomena and their relationships, classified and arranged. Illustrations from gravitation. . 5–9

CHAPTER II.

All physical science the result of observation. Inadequate ideas of this doctrine. Illustration; marble; sources and means of our knowledge of this substance. One species, or kind, of knowledge, not deducible from another, independent of observation. Optics. All the properties of light ascertained exclusively by observation. Functions of mathematical calculations. Functions of *à priori* reasoning. Newton. Fresnel. . . . 10–25

CHAPTER III.

All true relationships invariable. Error of the common saying, that *the exception proves the rule*. Nature and constitution of laws, or principles, of science. They consist, exclusively, in constant phenomena and relationships, clas-[vii]sified and arranged. Never in anything lying back of these phenomena and relationships. Gravitation. Chemical Science. The law of definite proportions. What it is. Other illustrations. Electricity. Light. . . 26–32

CHAPTER IV.

Most of our knowledge incomplete. Natural wish to render it perfect and absolute. Attempts and efforts to accomplish this end give birth to hypotheses. Nature and constitution of hypotheses. Their true relation to science. All science independent of hypotheses. Constitution of matter. The atomic theory of chemical combinations. Optics. Corpuscular and undulatory hypotheses. Newton's elastic ether. Uses and functions of theories. Their value overrated. Opinions of Newton and Davy. 32–52

CHAPTER V.

Arrangement and classification of phenomena and relationships. Principles and grounds of this arrangement. Illustrations. Marble. . 52–55

PART II.
PHILOSOPHY OF MEDICAL SCIENCE

PRIMARY PROPOSITIONS, ... 59, 60

CHAPTER I.

Definitions. Anatomy; Topographical; General; Microscopic; Chemical; Comparative. Physiology. Pathology. Etiology. Therapeutics. . 61–67
[viii]

CHAPTER II.

General prevalence of false notions. Medical science consists, exclusively, in the phenomena and relationships of life, classified and arranged. Anatomy. Physiology. Illustrations. Germination of seeds. Conditions of germination. Phenomena of germination. Respiration; its phenomena. . 67–74

CHAPTER III.

Extent of erroneous notions. . . . 75–76

CHAPTER IV.

Our knowledge of anatomy not dependent upon our knowledge of other branches of medical science. Our knowledge of one branch of anatomy does not include the knowledge of any other branch. . . 77–78

CHAPTER V.

Our knowledge of physiology not deducible from our knowledge of anatomy. Qualifications. Final causes. Illustrations. Brain. Stomach. . 79–85

CHAPTER VI.

Our knowledge of pathology not deducible from our knowledge of physiology. Qualifications. Illustrations. Inflammation. Differences in the susceptibilities of different organs to this process. These differences not to be accounted for on physiological grounds. Gastritis. Other diseases. . 85–100

CHAPTER VII.

Relations of pathology to its causes. Etiology. Our knowledge of the causes of disease, the exclusive result of ob-[ix]servation. Etiology not to be deduced from pathology. Illustrations. Age. Sex. Season. . . 100–102

CHAPTER VIII.

Relations of pathology to its modifiers. Therapeutics. Rationalists. Empirics. Therapeutics not deducible from pathology. Inflammation. Periodical diseases. Cinchona and arsenic; Relations between them. Action of remedies on disease, not deducible from their action in health. Opium. Cinchona. Calomel. Action of remedies on the human body, not deducible from their action on those of other animals. 103–120

CHAPTER IX.

Diagnosis; its importance, and its relations to therapeutics. Illustrations. Pleurisy; Typhoid Fever. . . . 121–127

CHAPTER X.

Diagnosis, twofold. — Nosological and Therapeutical. Elements and means of nosological diagnosis. Diseases not to be required to be wholly unlike each other. *Typhoidal* fever, and *congestion*, common elements. Locality of disease. Nature, or character of disease. Combination and succession of certain phenomena. Symptoms. Relative value of these several elements. Tendencies of modern researches. Therapeutical diagnosis. . . 127–146

CHAPTER XI.

The character and conditions of principles in medical science. These principles approximate, and not absolute. This approximative character fixed and determinate. Its degree of fluctuation confined within certain limits. Illustrations. Proportion of sexes at birth. Law of great numbers. Calculation of probabilities. Laws, or principles, of therapeutics; their complexity; difficulty of ascertaining them. [x] Gavarret. Conditions of these laws. Facts must be comparable. True value of therapeutical experience. Mistaken notions. . . 147–179

CHAPTER XII.

The nature and value of what are called *Medical Doctrines*. Universal prevalence of medical hypotheses. Their bad influences. Methodism. Cullen's Theory of fever. Homœopathy: State of its principles. Standard by which they are to be tried. Evil effect of *Medical Doctrines* upon the minds of medical men, and upon the interests of medical science. Broussais: His *History of Chronic Inflammations*, and his *Examination of Medical Doctrines*. Sydenham. How far interpretation may be allowed. . . . 180–224

CHAPTER XIII.

American Medical Doctrines. Dr. Rush. Dr. Miller. Dr. John Esten Cooke. Dr. Gallup. Drs. Miner and Tully. Samuel Thompson. . 224–250

CHAPTER XIV.

The principles and conditions of nosological arrangements. These arrangements necessary. Classifications in botany. The artificial and natural methods. Are diseases legitimate objects of classification? Defects of nosological systems. Examples of natural groups, or families. Exanthemata. Fevers. Phlegmasiæ. Cancer and tubercle. Neuroses. Definitions. . 250–273

CHAPTER XV.

Relations of Vital and Chemical Forces. . . 273–282

CHAPTER XVI.

Future Prospects of medical science. Conclusion. Causes [xi] of the slow progress and imperfect state of medical science. Diagnosis must precede therapeutics. Reasons of this. Complexity of therapeutical relationships. Probable extent of our

power over disease. French medical observation. British medical observation. American medical observation. . . 282–310
[xii]

PART FIRST

THE

PHILOSOPHY OF PHYSICAL SCIENCE [1]

"Non excogitandum est quid natura faciat, sed inveniendum." *Bacon.*

"The construction of the world, the magnitude and nature of the bodies contained in it, are not to be investigated by reasoning, which was done by the ancients, but to be apprehended by the senses, and collected from the things themselves. . . . They who before us have inquired concerning the construction of this world, and of the things which it contains, seem indeed to have prosecuted their examination with protracted vigils and great labor, *but have never looked at it.* . . . For, as it were, attempting to rival God in wisdom, and venturing to seek for the principles and causes of the world by the light of their own reason, *and thinking they had found what they had only invented,* they made an arbitrary world of their own. . . . We, then, not relying on ourselves, and of a duller intellect than they, propose to ourselves to turn our regards to the world itself and its parts." *Bernardinus Telesius. Quoted by Whewell. Philosophy of the Inductive Sciences.* Vol. II. p. 354.

"Itaque hominum intellectui non plumæ addendæ, sed plumbum potius et pondera; ut cohibeant omnem saltum et volatum. Atque hoc adhuc factum non est; cum vero factum fuerit, melius de scientiis sperare licebit." *Bacon, Nov. Org.* Lib. 1, Aph. CIV.

"Les hommes ne s' attachent aux faits qu' après avoir epuisé les hypothèses." *Broussais.*

"Whatever is not deduced from the phenomena is to be termed hypothesis and hypotheses, whether metaphysical or physical, or occult causes, or mechanical, have no place in experimental philosophy." *Sir Isaac Newton.*

"And if what I have said is but intelligible and true, and carries so much conviction with it, of its being so, that it may induce some others to pursue those methods of improving medicinal knowledge, which are herein recommended; or if it contains anything that is either useful or new, which may contribute something to its improvement, or may be the means of exciting some other physicians to make any farther discoveries or improvements in the medical science, which may be useful to mankind, I shall not think my time and labor lost." *William Hillary.* [2]

PART FIRST

PHYSICAL SCIENCE

PRIMARY PROPOSITIONS

Proposition First. All physical science consists in ascertained facts, or phenomena, or events; with their relations to other facts, or phenomena, or events; the whole classified, and arranged.

Proposition Second. These facts, phenomena, and events, with their relations, can be ascertained only in one way; and that is by observation, or experience. They cannot be deduced or inferred, from any other facts, phenomena, events, or relationships, by any process of reasoning, independent of observation, or experience.

Proposition Third. A law, or principle, of physical science consists in a rigorous, and absolute generalization of these facts, phenomena, events, and relationships; and in nothing else. It is identical with the universality of a phenomenon, or the invariableness of a relationship.

Proposition Fourth. A hypothesis is an attempted explanation, or interpretation, of these ascertained phenomena, and relationships; and it is nothing else. It consists in an assumption, or a supposition, of certain other unascertained, and unknown phenomena, or relationships. It does not constitute an essential element of science. All science is absolutely independent of hypothesis. [3]

Proposition Fifth. Theory is one of two things, according to the manner in which the word has been used. It is either a generalization of phenomena, and relationships, and in this case, identical with a law, or principle, of science; or, it is an attempted explanation of phenomena, and relationships, through the intervention of other assumed, and unascertained, phenomena, and relationships, and, in this case, identical with hypothesis.

Proposition Sixth. All classification, or arrangement, depends upon, and consists in, the identity, or similarity, amongst themselves of certain groups of phenomena, or relationships; and their dissimilarity to other groups of phenomena, or relationships. All classifications or arrangements are natural and perfect just in proportion to the number, the importance, and the degree of these similarities, and dissimilarities. [4]

PART FIRST

THE PHILOSOPHY OF PHYSICAL SCIENCE

CHAPTER I.

PROPOSITION FIRST

ALL PHYSICAL SCIENCE CONSISTS IN ASCERTAINED FACTS, OR PHENOMENA, OR EVENTS; WITH THEIR RELATIONS TO OTHER FACTS, OR PHENOMENA, OR EVENTS; THE WHOLE CLASSIFIED AND ARRANGED.

Object of this essay. All science consists exclusively in phenomena and their relationships, classified and arranged. Illustrations from gravitation.

THE sole object of this essay is an exposition of what I conceive to be the true principles of *medical* philosophy; and it will be mostly made up of this direct exposition. But, in order that I may be enabled to accomplish this object, successfully and satisfactorily, I have thought it necessary to state, in the first place, what I conceive to be the true, fundamental doctrines of the philosophy of all *physical* science. This I have done in the foregoing first, second, third, fourth, and sixth primary propositions. These doctrines, with certain modi-[5]fications, in one or two particulars, are identical with the true doctrines of the philosophy of medical science. There is no essential difference between the philosophy of physical, and that of physiological science. It happens, however, for reasons which it is not necessary here to give, that the former philosophy is susceptible of being rendered plainer, and more clearly intelligible, to most minds, than the latter.[1] This circumstance induces me to make use of the illustrations, which may be derived from a brief exposition of the true principles of the philosophy of physical science, as an introduction to the more important and principal work before me, the statement and exposition of the true principles of the philosophy of medical science. I do not think, that I can well and entirely accomplish the latter, without the aid of the former. At any rate, there are no other collateral sources, from which so important and so various assistance can be derived, as from these, which I have thus indicated; and for these reasons, I shall devote this, the first part of my essay, to this subject.

The first proposition, that which stands at the head of this chapter, does not require much illustration. Its truth is so manifest, as hardly to admit of any doubt. It would seem almost impossible, that there should be any difference of opinion in regard to its soundness, or any obscurity in its conception. I believe, nonetheless, it is true, that there has always been, and that there still is, in [6] the minds of most men, and in those of philosophical thinkers, a somewhat imperfect, or confused, apprehension of its doctrines. I do not think that its truth is seen and felt, as it should be, in the simplicity, the purity, and the absoluteness, which belong to it. The confusion, to which I allude, is this. There seems to be a common feeling, that the facts, phenomena, and events, with their relationships, classified and arranged, constitute, not the entire science, to which they belong, but only the *foundation* of the science. There is a feeling, that these facts and relations are to be used as elements, out of which, the science is to be built up, or constructed, by what is called *inductive reasoning*. The feeling implies, and the avowed doctrine growing out of it often asserts, that the science *is in this subsequent process of reasoning*, and not in the facts, themselves, and their relationships. We are constantly told, that the facts are to be used *as materials*, to be sure; that it is not safe to take for our materials anything but facts; that they constitute the *basis* of every science; but, after all this, the essential condition and constituent of the science is often placed, more in the process of reasoning, as it is called, than in the facts and their relationships. Now, what I wish to insist upon is this; that the science *is in the facts and their relationships, classified and arranged, and in nothing else.* The ascertained facts and their relationships, classified and arranged, constitute, in themselves, and alone, the science, and [7] the whole science, to which they belong. The science, thus constituted, is, so far, complete. No process of inductive reasoning, or of any other reasoning, no act of the mind, can add anything to what has already been done. The only reasoning, that has anything to do with the matter, consists, simply, *in the act of arranging and classifying* the phenomena, and their relationships, according to their differences, their resemblances, or their identity. Words are things; and I cannot doubt, that much obscurity and confusion would be removed from our conceptions of the nature of the philosophy of science, if this long-abused term, *inductive reasoning*, could be suffered to disappear from the language of science and philosophy; and if, for the indefinite and shadowy ideas, which it so often expresses, or attempts to express, could be substituted those, which are so clearly and obviously contained in this phraseology, — *the classification and arrangement of phenomena and their relationships.* [2]

In seeking for illustrations of the true nature of the philosophy of physical science, we turn, almost instinctively, first, to the phenomena of gravitation. These phenomena are the results of one of the simplest of all known relationships, — that of different portions of matter to each other, through space. This relationship is not mixed up with any others; it is not liable to be disturbed, or affected, by any others. It has been very thoroughly [8] and fully investigated; and we have every reason to believe, that our knowledge of it is as absolute and complete, as human knowledge is capable of becoming. Now, the whole science of gravitation consists in its phenomena, classified and arranged, and in nothing else. These phenomena, thus classified, constitute, *not the foundation, and the material, merely*, on which, and out of which, by some recondite process of intellectual powers, called inductive reasoning, the science is to be constructed; they *are*, the science, *in themselves, wholly, and absolutely.*[3] When all the phenomena, depending upon this single relationship of matter, have been ascertained, and classified, the science of gravitation *is complete*; it is finished; there is nothing more to be done. Nothing can be added to it be any subsequent process of reasoning, or act of the mind. And the same thing is true of all the departments of physical science; but inasmuch as this subject will necessarily receive further incidental illustration, in other parts of my essay, it is not important that I should dwell upon it any longer, for the present. [9]

CHAPTER II.

PROPOSITION SECOND

THE FACTS, PHENOMENA, AND EVENTS, WITH THEIR RELATIONSHIPS, CLASSIFIED AND ARRANGED, CONSTITUTING PHYSICAL SCIENCE, CAN BE ASCERTAINED ONLY IN ONE WAY; AND THAT IS BY OBSERVATION, OR EXPERIENCE. THEY CANNOT BE DEDUCED OR INFERRED, FROM ANY OTHER FACTS, PHENOMENA, EVENTS, OR RELATIONSHIPS, BY ANY PROCESS OF REASONING, INDEPENDENT OF OBSERVATION, OR EXPERIENCE.

All physical science the result of observation. Inadequate ideas of this doctrine. Illustrations; marble; sources and means of our knowledge of this substance. One species, or kind, of knowledge, not deducible from another, independent of observation. Optics. All the properties of light ascertained exclusively by observation. Functions of mathematical calculations. Functions of *à priori* reasoning. Newton. Fresnel.

NOT only does all physical science consist, exclusively, in facts, phenomena, and events, with their relationships, classified and arranged; but these phenomena and relationships can be ascertained and classified in only one way; by only one method — that of observation. No single phenomenon, or property, or relationship of objects of physical science can be deduced, or inferred from any other phenomenon, or property, or relationship, unless the former is already contained in the latter. This independence of each separate class of phe-[10]nomena, and relationships, is entire and absolute. It is essential, to a clear comprehension of the philosophy of medical science, that this doctrine, thus stated, in its connexion with physical science, should be fully unfolded and distinctly seen; and to do this, is the object of the present chapter.

Ever since the time of Francis Bacon, the language of philosophy has been almost uniform upon this subject. The world has been constantly told, that all science, except that of a purely speculative, or metaphysical character, depends upon observation. This language has been eloquent and emphatic in its praises of the Baconian method of investigation; and it has been filled with warnings against the danger of what it calls speculative reasoning, and premature conclusions. But, notwithstanding all this, it is true, I think, even in physical science, — it is true, I know, in physiological science, — that the common conception of the doctrine of which I am speaking is inadequate and incomplete. The entire independence of each other, — so far as our knowledge of them is concerned, — of the several classes of phenomena and relationships, which go to make up physical science, is only partially and imperfectly comprehended. The dependence of our knowledge of each and every class of phenomena and relationships, upon direct observation of the particular class itself, is more exclusive and absolute, than seems to be generally supposed. It is true of this doctrine, as it is of that contained in [11] my first proposition, that *ideas*, and *reasoning*, and *deduction* are supposed to have much more to do with it, than is really the case. There is a common feeling, that such a connexion has been established between these different classes of phenomena and relationships, at least in many instances, as to enable us, one class having been already ascertained by direct observation, to infer, or deduce the existence of the others, by some act of the mind, independent of further and direct observation. This misconception, if such is its character, I wish now to expose and remove.

For the purpose of illustration, let us take, in the first place, any one of the common forms of inorganic matter, as they exist about us. What is true of one of these forms, so far as my present object is concerned, is true of all the others; and amongst these substances, there is no one better adapted, on the whole, to the end which I have in view, than that which is known by the name of marble.

Our first and most readily acquired knowledge of this substance has reference to the phenomena which it presents in its direct relation to our senses. These phenomena, in this relation, constitute what are called its manifest, sensible properties. They consist of its color, varying in its different varieties; its weight, or specific gravity; its hardness; its brittleness; its mode of fracture; its granular, or crystalline structure; its elasticity, and its susceptibility to polish. Our [12] knowledge of each and all of these obvious, physical properties is the result of direct observation of the particular, individual property itself; and the existence of no one amongst them could have been inferred, or deduced by any conceivable process of reasoning, independent of observation or experience, from the presence of any one, or more, of the others.

Another element, in our knowledge of marble has reference to its intimate composition; we are able to ascertain the number, the character, and the relative proportions of the simple, elementary substances, which are united to constitute it what it is. We find that it is composed of two substances, carbonic acid, and lime, and that these substances are united in definite and fixed proportions., ascertained by weight. We then find, on further examination, that the carbonic acid is, itself, composed of two substances, carbon, and oxygen, united in definite and fixed proportions; and, also, that the lime, like the acid, is composed of two substances, calcium and oxygen, united, also, in definite and fixed proportions. The oxygen, the carbon, and the calcium, are, in the present state of science and art, not susceptible of further division, or analysis. The union, in certain proportions, of these three elementary substances, constitutes the chemical composition of marble. Our knowledge of this composition is obtained through the agency of chemistry; and it is exclusively the result of what may be called chemical [13] observation. It could never have been acquired through any other means, or from any other sources. Certainly, there is nothing in the sensible qualities of marble, which could have indicated, in the remotest degree, the character of its intimate composition; and no acquaintance, however perfect, with the separate properties of the several elementary substances, themselves, of which it is composed, could have enabled us, by any process of reasoning, to infer or deduce the result of their combination, in the production of marble itself.

Our knowledge of marble is completed, when we have ascertained, in addition to the foregoing properties, its various relations. The most obvious of these relations is that of geographical locality, — the distribution of marble throughout the various regions of the earth. Another, somewhat analogous to this, refers to its position amongst the several layers, or strata of substances, which are more or less regularly arranged,[1] one above the other, to form the solid crust of the globe. This position, with the circumstances attending it, constitutes its geological relations. By the chemical relations of marble, I do not mean all the relations of its ultimate, elementary constituents; although it may, properly enough, be said, that our knowledge of marble is absolute and complete, other things being equal, in proportion to the extent and accu-[14]racy of our knowledge of these relations. That is, the more extensive and complete our knowledge of all the properties and relations of carbon, oxygen and calcium is; in a certain sense, at least, the more extensive and complete is our knowledge of marble itself. But, strictly speaking, the chemical relations of marble can hardly said to be coextensive with the relations of its elementary constituents. Still, they are numerous and interesting. For instance, there are several substances, amongst the solid materials constituting the crust of the earth, which differ, more or less widely, in their manifest, sensible properties, and in some of their relations, from marble; which have, notwithstanding these differences, precisely the same chemical composition. One of these substances is lime-stone; another is chalk; and a third is marl. All these substances are, like marble, carbonates of lime. Again, there are other substances, identical with the foregoing ones in their chemical composition, but differing widely from them in nearly all their other properties and relations. Amongst these it is sufficient for my present purpose to mention Iceland spar. This is a carbonate of lime, like marble, and like chalk; but it differs from these substances in its hardness; in its foliated fracture; in its structure, and, especially in its singular relations to light and electricity. In the language of optics, it is doubly refractive, which the others are not.

Another relationship of marble, is that of its [15] particles to each other, and to those of all other matter, through space. A piece of marble, when elevated to any distance, great or small, from the surface of the earth, and then left to itself, immediately falls to the surface; and the velocity of its motion increases in a uniform ratio, which ratio has been accurately ascertained. Furthermore, all other material substances are the subjects of the same phenomena; under the same circumstances, they all fall, with the same uniform and increasing rations of velocity, to the earth. In addition to this, these substances not only tend towards the earth, and that with a certain force, which can be measured, but they all tend towards each other, and this also, with a certain force, which can be measured. This universal fact, or phenomenon, is expressed by the term gravity, or gravitation, or the attraction of gravitation.[2]

The same piece of marble, when not falling, and when left to itself, will remain in the same position; and this is also true of all other material substances. This property of matter is expressed by the term *inertia.* This quiescence, or rest, in the same position, can only be destroyed by the force, or pressure, of some other body, acting upon that at rest. The directions, the velocities, and so on, of these imparted changes of place, with their relations to the tendencies, which these same moving bodies have to approach each other, and all other material bodies, constitute the elements of the science of motion. [16]

Such, then, are at least the principle relationships of marble. They are, as is the case with all other substances, numerous and interesting. But, here, again, as in the instance of its sensible qualities, and its chemical composition, there is no such connexion between them, as to enable us, independent of observation, to ascertain the existence of any one of them, from the presence of any other. Each distinct and peculiar relationship can be ascertained in one only way, by one only method, — that of observation of the relationship itself. The presence of any one relationship does not imply, or involve, the presence of any other. Our knowledge of the relations of marble to light could never have been derived, by an act of the pure reason, inductive, or otherwise, from our knowledge of its relations to electricity, or to heat, or to other bodies through space; and so on of all its relationships. No one of these is contained, or included, in any of the others, and is not, therefore, susceptible of being deduced from them.

And this doctrine is universal in its application; it is true of all the properties, all the phenomena, all the relationships, of all substances. I have already spoken of the independent character of our knowledge of the chemical constitution and relationships of marble. Our knowledge of the like constitution and relationships of all other substances is equally independent; it is exclusively derived from observation of the constitution, and the relationships themselves. Is there anything in [17] the sensible properties of water — is there anything in its dynamic relations — in its tendencies toward other bodies — in its inertia, or in its motions, which could, in any conceivable way, have led us to the knowledge, that it is composed of two simple substances, so different from itself, and from each other, as oxygen, and hydrogen? Is there anything in the other properties and relationships of these two substances, from which we could have inferred the production of water, by their combination in certain proportions? Most assuredly there is not.

In the further development of this doctrine, let us look, for a moment, at the manifold and beautiful relationships of light. Not in all physical science — not in astronomy itself — have there been any more wonderful achievements of human genius, than in optics. Nowhere, have the ingenious contrivances of art, and the nice applications of science, been productive of more marvellous and positive results, than here. But, in every instance, there results have been the fruit of simple observation, generalized, to be sure, and applied, as in the case of the phenomena of gravitation, by the aid of mathematical calculation. All the various relations of light to other substances; all its subtle and mysterious properties; its influence upon chemical combinations; the influence upon *it* of the intimate or molecular structure of bodies, through which it passes; the recondite affinities by which it is linked to heat, electricity, and gal-[18]vanism, have, each and all, been ascertained, so far as they are ascertained, solely and wholly, by simple observation, thus generalized, and applied; and by observation of each separate property and relationship. Certainly, there is no conceivable process of inductive reasoning, by which, the mind of Sir Isaac Newton[3] could have arrived at the knowledge of the heterogeneous and compound nature of light. It was with the prism, and his eyes, and not by any magic of his great intellect, that the web of its homogeneous rays was first unwoven and analyzed, and its composition ascertained. It was by means of the thermometer, and by this means alone, that Dr. Herschel[4] determined the presence, in the solar spectrum, of heating rays, independent of the rays of light. The original discovery of what is called the polarization of light, the development of which has led to such extraordinary results, was quite accidental even. It was revealed to M. Malus,[5] by a casual turn of the prism, through which, in 1808, he was gazing at a brilliant sunset, reflected from the windows of the Luxembourg palace, in Paris.

It has been said, I know, by the highest living authority, that some of the more abstruse and subtle properties and relations of light have been ascertained, and demonstrated, by pure *à priori* reasoning. Sir John Herschel speaks of the investigations and discoveries of Fresnel,[6] in connexion with the effects produced upon rays of light, by doubly refracting substances, as of this [19] character. These properties and these relations, thus supposed to have been ascertained, by means of pure reason, through the aid and instrumentality of mathematical calculations, of great length and complexity, are not sufficiently obvious and intelligible, to be used for my present purpose of popular illustration. But, it seems to me, that the functions of these mathematical processes have been mistaken, in the agency which has been thus assigned to them. It will be found, I believe, upon a close examination, and a strict analysis of these processes, and of the part which they play in optical science, that they are wholly incapable of being made the means of *discovering* any new property, or any new relationship of light. They are used for the purpose of illustrating, extending, and applying to new, but analogous, circumstances, certain phenomena and relations of light, already ascertained by observation. This is the province, it seems to me, and the only province; these are the functions, and the only functions, of such calculations, in all the physical sciences. And although I am not capable of fully comprehending them, I have no wish to seem, even, to detract from their importance. I am fully aware, that this importance is paramount; and that it cannot be exaggerated. I am fully, aware, that the science of optics owes very much of its perfection and beauty to the complex and difficult calculations of Newton, and Young[7], and Fresnel; and that the science of astronomy could [20] hardly be said to exist, independent of similar calculations. But in both these instances, and in all others, it seems to me, that the functions of these calculations consist, solely and exclusively, in the development, the generalization, the extension, and the application to new circumstances and conditions, of phenomena and relationships, previously ascertained by simple observation. The pure mathematics of Newton, La Grange,[8] and La Place,[9] constituted only an instrument, or apparatus, by whose subtle properties, and stupendous power, these philosophers were enabled to measure and estimate the force of gravity, in all conceivable conditions, and under all possible circumstances of difficulty and complexity. With all its wonderful subtlety, with all its stupendous power, it could no more *discover* a new property, or relation, of an atom of matter, than it could create the worlds, whose motions it so accurately measures, and whose relations to each other, it estimates with such consummate precision, and such marvellous skill. It might seem quite impossible, that the conclusions of Sir Isaac Newton, in regard to the dependence of the colors in the solar spectrum, upon different velocities in the motions of the assumed particles, constitution the several kinds of rays, should have been the result of mere observation. When, taking his researches for a foundation, it is alleged, on the undulatory hypothesis, that the red color of the spectrum is occasioned by the vibrations of the ethereal wave, [21] the length of which wave is equal to the 0'0000266th part of an inch, and the number of whose vibrations amounts to 458 millions of millions in a second, we may well be startled at the pretension, which profess to have estimated these numbers, and to have measure these velocities. It is, nevertheless, strictly true, that these almost transcendental results, so far as they are established, have been established, not by any high and refined processes of pure *à priori* reasoning, but by simple observation, generalized and applied, through the agency of mathematical calculation, as an instrument and means.

There is a certain sense in which many of the more subtle and recondite phenomena and relations of light, like those of which I have just spoken, and some others, may be said to be ascertained by induction, or inference. But even in these cases, we shall find, on a careful examination and analysis of our methods of investigation, that all our knowledge is the result of observation, and of observation alone. Thus it has been ascertained, that the intervention, under certain conditions, of very minute fibres or particles, between the eye and a luminous body, causes the body to be surrounded with a ring of colors; and that the width, or diameters of these rings increase with the size of the fibres, or particles, by the action of which they are produced. Dr. Young proposed and instrument, called an *eriometer*, founded upon the ascertained relationship be-[22]tween the size of the fibres, or particles, and that of their corresponding rings, round the luminous body, to be used for the purpose of ascertaining the size of these particles, themselves too minute for direct measurement, by ascertaining the width of their rings. Dr. Wollaston[10] found, by measurement, the diameter of the seed of the *lycoperson boviste* to be the 8500th part of an inch; and then by comparing with the rings, produced by this seed, those corresponding to other and much smaller particles of matter, he ascertained the size, or the diameters of these latter. There is no propriety in saying, that Dr. Wollaston *deduced*, or *inferred*, by any process of pure reasoning, the length of these diameters. He merely made use of the rings, produced by the action of the particles on light, as an instrument, or scale of measurement, wherewith to determine the diameter of the particles themselves, so minute as to be inappreciable by any other means. He converted them into an *eriometer*. By a beautiful application of the same instrument of observation, Sir David Brewster[11] ascertained the diameters and shape of the extremely delicate fibres of which the crystalline lens of the eye is composed. By a still more refined application of other known relations, and properties of light, the arrangement of the grooved surfaces of mother of pearl, and the internal structure of various crystallized bodies, beyond the powers of the microscope, have, to a certain extent, at least, been ascertained. It is [23] sufficiently obvious, I hope, that, in all these instances, we are indebted, for our knowledge, not to any intellectual process of induction or inference, or *à priori* reasoning; but to observation of each property and relationship, and to this alone.

There is another seeming qualification of the doctrine, that I am endeavoring to illustrate, about which it may be necessary to say a few words. It has often been alleged, for instance, that Sir Isaac Newton inferred, by a process of *à priori* reasoning, the combustibility of the diamond, before this combustibility had been demonstrated by observation. But what did Newton really do in this case? Manifestly this, and nothing more. A relationship had already been noticed between two certain properties, or phenomena, — at least in many bodies, to wit, their refractive power, and their combustibility. Newton's reasoning, as it is called, consisted, simply, in the suggestion, or conjecture, that this relationship might be absolute and universal; and, if so, that the diamond would prove to be combustible. The only reasoning in the case consisted in the application to new circumstances of an assumed relationship.[12] It has been said of Fresnel, that he "proved, by a most profound mathematical inquiry, *à priori*," the existence of certain subtle properties of polarized light. But here, again, what did Fresnel really do? He showed, by the agency of his mathematical calculations, that certain relationships of light, assumed, or ascertained by observation, in certain condi-[24]tions, must, if these relationships were true and genuine, exist, also, in all other identical conditions. He showed, that if certain modifications of light, wrought in its properties, by the action of Iceland spar, during its passage through this substance, were dependent upon certain peculiarities in its crystalline structure, then the same modifications must be produced in other substances, identical in these peculiarities of structure with the Iceland spar. He applied, merely, and generalized, by means of his calculations, a phenomenon, or relationship, of light, already ascertained by direct observation. [25]

CHAPTER III.

PROPOSITION THIRD

A LAW, OR PRINCIPLE, OF PHYSICAL SCIENCE CONSISTS IN A RIGOROUS, AND ABSOLUTE GENERALIZATION OF THE FACTS, PHENOMENA, EVENTS, AND RELATIONSHIPS, BY THE SUM OF WHICH, SCIENCE IS CONSTITUTED; AND IN NOTHING ELSE. IT IS IDENTICAL WITH THE UNIVERSALITY OF A PHENOMENON, OR THE INVARIABLENESS OF A RELATIONSHIP.

All true relationships invariable. Error of the common saying, that the exception proves the rule. Nature and constitution of laws, or principles, of science. They consist, exclusively, in constant phenomena, and relationships, classified and arranged. Never in anything lying back of these phenomena and relationships. Gravitation. Chemical Science. The law of definite proportions. What it is. Other illustrations. Electricity. Light.

ALL genuine and legitimate relationships are invariable and constant. This, indeed, is only another mode of stating the doctrine of the ancient axiom, that like causes, under like circumstances, must be followed by like effects.[a] An event, having once occurred, will always occur, under [26] the same circumstances; a phenomenon, having been once observed, will always be observed, in a like state of things; a relationship, once ascertained, will never fail, under the same condition of the related substances, or phenomena. This, at any rate, must be true so long as the present constitution of the universe continues.[1] If oxygen and hydrogen are united in certain proportions to constitute the drop of water, which holds in solution the coloring matter of the ink, wherewith these words are written, so do they unite in the same proportions to make up the waters of the ocean and the rivers. The rays of light, now falling upon this page, have occupied precisely the same period of time, on their journey hither, from their great source and fountain, as was occupied by the first which visited the earth, when the sun was set in the firmament. Those which fell upon the seas, on the morning when the waters were first "gathered together unto one place," were changed from their direction at the same angle, that now marks their divergence. All exceptions, as they have been called, to this invariableness and uniformity are apparent only, and not real. They are the result, only, of our imperfect knowledge. The old saying, so constantly and so blindly repeated, that the exception proves the rule, is as destitute of truth, as it is of

---

[a] Whether our belief in the truth of this doctrine depends, in any degree, upon experience, or wholly upon an innate and fundamental property of our mental constitution, it in no way concerns my present purpose to inquire. It is sufficient for me, that this *idea of cause*, is, in the words of Professor Whewell, "an indestructible conviction, belonging to man's speculative nature."

meaning. Such an exception can prove only one thing, and that is, that the rule is not fully understood, or completely ascertained. The relations of many substances [27] and agencies, in nature, to other substances and agencies, are so numerous, and so complex; they so cross, and intermingle with, and modify each other, as to render their analysis, with our imperfect means of investigation, often difficult, and sometimes impossible. But even in these instances of combination and complexity, we should find, if our means of investigation were adequate to their analysis, and separation from each other, that each single series of legitimate relationships is as absolute and constant, as that simples and sublimest of all, which directs a falling apple to the earth, and guides the heavenly bodies, in their circuits through the celestial spaces. Without this constancy and uniformity, there could be no such thing as what we call a principle, or law of science; there could, indeed, be no such thing as science. "Order is heaven's first law;" and the essential condition of all order rests in this fundamental and absolute fact of the uniform constancy of phenomena, and the fixed invariableness of relationships, under the same circumstances.[b] [28]

My object, in this chapter is to show, that all laws, or principles, of science consist, merely, in these constant and invariable phenomena and relationships. This is necessary, because there is a feeling, more or less common, that a law, or principle, of physical science is something more than a universal fact, or a uniform relationship; and that it consists in some unknown power, or agency, lying back of the phenomena, or interposed between those which are related to each other; of which power, or agency, the phenomena, themselves, are only the manifestation, and the result.[2] To illustrate my meaning, let us first take what is called the law, or principle, of gravitation. This law consists in the generalization of a single ascertained phenomenon; it is the expression of a single, universal fact, to wit: that all substances, with the exception of the few, which are called imponderable, when left to themselves, and not restrained, or prevented, by any counteracting, or opposing, forces, will approach each other; and this in a certain ratio of velocity, which is susceptible of admeasurement. The law consists simply in this generalization, and in nothing else; the principle is the expression of this [29] fact, and of nothing else. The universality of the fact, or the generalization of the phenomenon, constitutes the sole element of the law. One expression is, literally and absolutely, equivalent to the other. No new element can be introduced into the law, by the super-addition of other ideas. The supposition of the existence, between

---

[b] Professor Whewell says, that no law, or proposition, absolute and universal in its character, can be established by observation, or experience alone; for the reason, that experience is limited, and not commensurate with the law or proposition to which it refers; that the laws, for instance, of gravitation, light, and so on, so far as they are established by observation alone, are known to be general only, and not universal; and that they acquire the stamp and character of universality only by the light shed upon them by the fundamental ideas of the mind. But, certainly, the doctrine of the absolute invariableness of all true relationships, of the fixed uniformity of the phenomena of nature, a doctrine universally and necessarily admitted, gives to the laws ascertained by observation the same degree of positiveness, as that which belongs to any conceivable laws whatever. I do not see how they are any more *contingent*, than those with which Professor Whewell contrasts them, and which he calls necessary laws or truth. *Phil. Ind. Sci.* vol. i. p. 61.

the bodies, tending towards each other, of some invisible and inappreciable force, or power, or agency, in the form of an ether, or in any other supposable form, would, even if the reality of the force were demonstrated, in no way affect the truth of what I have said. The relationship might thus be rendered less direct, and simple, by this intervention of a new phenomenon, or series of phenomena; but the law, or principle, itself, would still remain precisely what it now is — the expression of a universal fact — and nothing else. The essence of Newton's immortal discovery consisted in seeing and demonstrating the absolute simplicity, universality, and invariableness of this great relationship; and his dynamical system of the universe consists in its development and application.

The same kind of illustration may be applied to any and to all the laws, or principles, as they are called, of physical science, and with the same results. One of the fundamental principles of chemical science is this, — that different bodies combine with each other in definite proportions, ascertained by weight. The law is in this uni-[30]versal fact, and not in any other conditions, or circumstances, that may be supposed to attend it. The principle, or the law, and the expression of this simple fact, are precisely identical. No single idea enters into one, that does not equally enter into the other. It is a law of electrical science, that the two kinds of electricity, — the positive and the negative, as they are called, — are always evolved in equal quantities; — that there cannot be an evolution of one, without an exactly corresponding, equivalent evolution of the other. It is a law of optical science, that light, in passing obliquely from a rarer into a denser medium, is turned, at a certain angle, depending upon the degree of difference in the density of the two media, towards a line perpendicular to the surface of the denser medium which it enters. In these, as in the foregoing, and in all other instances, the law, or the principle, is constituted, exclusively, by a rigorous and absolute generalization of the phenomena, or the relationships, which are its subjects. There is no other element than this, entering into the constitution of the law. The law is absolute, just in proportion to the universality of the phenomenon, or the invariableness of the relationship; and just so far as these are not rigorously and positively established, is the law partial and incomplete. Every separate and individual phenomenon, every separate and individual relationship, constitutes an element in a law or principle of science. There are just as [31] many of these separate and independent laws, or principles, as there are distinct classes of phenomena, or relationships.

CHAPTER IV.

PROPOSITION FOURTH

A HYPOTHESIS IS AN ATTEMPTED EXPLANATION, OR INTERPRETATION, OF THESE ASCERTAINED PHENOMENA, AND

RELATIONSHIPS, CONSTITUTING SCIENCE; AND IT IS NOTHING ELSE. IT CONSISTS IN AN ASSUMPTION, OR A SUPPOSITION, OF CERTAIN OTHER UNASCERTAINED, AND UNKNOWN PHENOMENA, OR RELATIONSHIPS. IT DOES NOT CONSTITUTE AND ESSENTIAL ELEMENT OF SCIENCE. ALL SCIENCE IS ABSOLUTELY INDEPENDENT OF HYPOTHESIS.

Most of our knowledge incomplete. Natural wish to render it perfect and absolute. Attempts and efforts to accomplish this end give birth to hypotheses. Nature and constitution of hypotheses. Their true relation to science. All science independent of hypotheses. Constitution of matter. The atomic theory of chemical combinations. Optics. Corpuscular and undulatory hypotheses. Newton's elastic ether. Uses and functions of theories. Their value overrated. Opinions of Newton and Davy.

OUR knowledge of nearly all the properties, phenomena, and relations, of the substances and agencies, which constitute the objects of physical science, is partial and imperfect. It is very rarely, if ever, absolute and complete. The senses, even [32] when aided by all the means and appliances of science and art reveal to us only a part, and probably a small part, of the properties, phenomena, and relations, of the substances and agencies, which go to make up the material universe. Behind and beyond all these appreciable properties, phenomena, and relations, we feel that there must be others, with which these are connected, and upon which they depend. We feel that the position which we occupy, is at the confluence of numberless infinities, ourselves walled in, on every side, with impenetrable darkness, into which darkness, and from which, these infinities flow. The restless and inquisitive mind, from its very constitution insatiable, and ever unsatisfied with its actual and absolute possessions, endeavors to imagine the phenomena, which it cannot demonstrate; it struggles to overleap the boundary, whose inexorable circumference cages it in; and, failing to do this, it fills the infinite and unknown regions, beyond and without it, with its own creations. The fruits of these efforts, the results of these struggles, and of this constitution of the mind, are theories and hypotheses; or, in other words, interpretations and explanations, of appreciable and ascertained properties, phenomena, and relationships, through the medium of other unknown or imagined properties, phenomena, and relationships. It is the object of this chapter, to point out the true character of hypotheses, or the-[33]ories, and to show the nature of their connexion with physical science.[1]

Amongst the earliest physical hypotheses, were those which had reference to the intimate and ultimate constitution of matter. No region could be opened to the discursive and speculative disposition of the human mind, so captivating and so boundless as this; and we accordingly find, that all philosophies, from the pure and subtilized idealism of Plato, even to the stern and triumphant generalizations of Newton, have allied themselves, more or less closely, to some hypothesis of this character. They have thus endeavored to explain the appreciable composition and properties of material substances, by supposing these substances to consist of certain ultimate atoms, which atoms they have endowed with definite qualities and attributes. In the same spirit, we still continue to say, that these atoms are solid, indivisible, impenetrable, and so on. We talk about their shape, their weight, their hardness, their number, and the spaces by which they are separated from each other. We fill up these spaces with electrical matter, or with some other ethereal fluid, of almost infinite subtlety; and then we go on to deduce many of the obvious properties of matter, from supposed relations between the particles of this fluid amongst themselves, and from other supposed relations between this fluid and the atoms of matter.[2] Now, what I wish particularly to insist upon is this, — that all these as-[34] sumed phenomena and conditions are altogether matters of pure supposition. They enter, in no way, into legitimate science, so far as the properties and relations of matter are concerned; they do not constitute one of its elements. Physical science is wholly and absolutely independent of them. The very existence of ultimate molecules, or atoms, with the qualities which we so confidently assign to them, is a matter of the purest conjecture; it is entirely a fiction of the mind. They may, or they may not, exist in nature. And I may remark, further, that this utter and absolute ignorance in which we are placed, of the ultimate constitution of matter, and of the relations which may exist between its elementary constituents, ought at least to teach us caution in the construction of theories, or hypotheses, founded on an assumed condition of this constitution, and of these relations, and modesty in the promulgation and defence of such theories, or hypotheses. Art, with its manifold appliances, and science, with its marvellous insight, have opened to us so many of the mysteries of matter, that we are in danger of forgetting how infinite the distance may still be between what is known, and what is unknown. The smallest visible particle of marble appears, under the microscope, like a huge irregularly-shaped block; and it may be, that the minutest atom, which is revealed to our straining sight, only by the most powerful microscope in the strongest light, contains, still far within its appreciable [35] form, the structure and arrangement upon which its properties depend. It may be, that infinitely beyond the boundaries of this microscopic vision, all those processes are carried on, and those relations are established, which constitute the particle of matter what it is. Far, far beyond this visible boundary, and hidden within unapproachable recesses, actions may be going on, between the ultimate constituents of matter, not only utterly removed from our knowledge, but as truly beyond our powers of conception even, as eternity and space are beyond our powers of measurement, or estimate. Lest the tone of these remarks should seem exaggerated, I will quote the words of Professor Whewell[3] upon this subject, with which I have become acquainted since my own were written. "But when we would assert this theory," he says, — of ultimate particles, — "not as a convenient hypothesis for the expression or calculation of the laws of nature, but as a philosophical truth, respecting the constitution of the universe, we find ourselves checked by difficulties of reasoning, which we cannot overcome, as well as by conflicting phenomena, which we cannot reconcile."[a]

---

[a] Phil. Ind. Sci. vol. i p. 414.

Observation has shown, that when different bodies unite *chemically*, as we term it, so that a substance differing in its properties and relations from those of its component or constituent ele-[36]ments, is formed, they unite in certain fixed and determinate proportions. In order *to account for* this general fact, as a means of *explaining* and *interpreting* this law of combination, we resort to the assumed atomic constitution of matter, of which I have just been speaking; and we suppose, that a single atom of one substance, or element, can unite only with a single atom, or with two, or three atoms, and so on, of another substance or element. Now, all this again is a matter of pure supposition. The very existence, as I have already said, of the atoms themselves, with the properties that are ascribed to them, is wholly conjectural; and their union with each other, according to the Daltonian theory, is equally so. However plausible and beautiful this theory may now be considered, it is quite possible, that new and widely different explanation of the general fact, or law, of combination in definite proportion, may yet be suggested, displacing the other, and removing it entirely from the province of chemical science.[4] Professor Whewell says, — "So far as the assumption of such atoms as we have spoken of, serves to express those laws of chemical composition which we have referred to, it is a clear and useful generalization. But if the atomic theory be put forward, as asserting that chemical elements are really composed of atoms, that is, of such particles no longer divisible, we cannot avoid remarking, that for such a conclusion, chemical research has not afforded, nor can afford, [37] any satisfactory evidence whatever." At any rate it is true, that the science of chemistry is wholly independent of this, and of all other interpretations of its phenomena; these interpretations do not constitute any of its essential or legitimate elements.

There is, perhaps, no department of physical science, in which theory, or hypothesis, has played a more prominent part, than it has in optics. If the mind of Newton was unable to rest satisfied with the simple establishment of the laws of gravitation; if he found it difficult to conceive, that the atoms and the masses of the universe should tend towards each other, without the intervening agency of some material bond of union, still more difficult was it for the same mind, to be satisfied and content with the discovery of the appreciable properties and phenomena of light. Many of these properties and phenomena appeal so strongly and directly to one of the most positive and accurate of our senses; they are of such wonderful and multiform variety and beauty, that by an instinctive and irresistible impulse of the mind, we refer them to other and more remote phenomena, with which we suppose them to be connected, and upon which we suppose them to depend. Newton supposed, accordingly, that light consisted of very minute particles, of a peculiar imponderable matter, given off, principally, from the surfaces of all self-luminous bodies; the various motions, combinations, and relations of which particles, gave [38] rise to all the phenomena of light. The existence of these particles could, in no way, be demonstrated; their existence and properties were assumed, as the most convenient and plausible means of accounting for and explaining the appreciable properties and phenomena of light; and this assumption, with its development, constituted what has been called the material, or corpuscular theory of light. The progress of optical science, subsequent to the great discoveries of Newton, revealed the existence of properties and phenomena, which his hypothesis was inadequate satisfactorily to explain; and another theory, contemporaneous in its origin, or nearly so, with that of Newton, is now, very generally at least, adopted in its stead. This latter theory assumes the existence, in all space, between the masses and the atoms of matter, of a subtle and elastic ether, upon the vibratory motion of the particles of which, all the phenomena of light are supposed to depend; and this assumption constitutes what has been called the ethereal, or undulatory theory of light. I may remark here, that the latter is as much a corpuscular, or material theory, as that of Newton; both theories assuming the existence of material particles, in the motion of which the phenomena of light are supposed to consist.

Here, as in all the preceding illustrations, I wish it to be seen, that the theory, or hypothesis, is merely a mode of explaining and interpreting, or rather [39] of attempting to explain and interpret, certain ascertained phenomena and relations, by the assumption, or supposition, of the existence of unascertained and unknown phenomena. Certainly, the science of optics does not consist in either of the above theories. No single individual ever made so many and so brilliant discoveries in this science, as were made by Newton; and this may be alleged without obscuring one ray of the halos which may be said, almost literally, to surround the names of Malus, of Frauenhofer,[5] of Fresnel, of Young, of Herschel, and of Brewster; but the theory which Newton adopted, in order to explain the properties and phenomena which he discovered, is now almost universally rejected; it is regarded as insufficient, or erroneous. But these properties and phenomena, constituting, so far as they go, the science of optics, are altogether unaffected by the rejection of the hypothesis which assumed to explain them. The fate that has befallen the theory of Newton, may yet also befall that of Huyghens[6] and Young. In the infinite future, which will ever stretch out before the advancing progress of science and art, properties and relations of all forms of matter, no unimagined and undreamed of, may yet be discovered, by means and processes of investigation now wholly hidden, which shall utterly overthrow the present theory of light, beautiful and stable as it appears to be. Or if this does not happen, another result is likely to follow, which comes to [40] much the same thing; and this result, there can hardly be any presumption in saying, so far as the recent progress and the present state of science can enable us to conjecture, is the most probable of the two. The existence of the supposed ether, certain elements of its constitution and of its relations, now only inferred, or deduced, from the phenomena of light, may yet be positively ascertained. In this case, the theory is no longer a theory; the hypothesis is no longer a hypothesis. Their character is destroyed. The phenomena are no longer assumed; and they take their place amongst the other known phenomena of the science, constituting now one of its permanent and legitimate elements. If there is still need of theory, or hypothesis, it must be placed one step further back, still beyond that wall of darkness, which has only receded, instead of having been destroyed. The new theory must consist in other assumed properties and relations, assumed for the purpose of explaining those, now ascertained and demonstrated, of the particles of ether. And so must it ever be. Now and always — in optics and in all other sciences — the science itself consists in ascertained phenomena and relations; hypothesis, or theory, in other assumed phenomena and relations, — assumed as convenient or plausible means of explaining, or accounting for, these.

There is one very common feeling in regard to these interpretations and explanations, which is, [41] that they render the phenomena, to which they are applied, more intelligible — more easily comprehended and understood, than they would otherwise be. It seems to me, that there is some fallacy in this feeling, or, at least, that its alleged value is exaggerated. We shall find, I think, on a close examination of the matter, that the difficulty to which I refer, is only changed in the place which it occupies, by these explanations; that it is neither removed, nor very materially diminished, by them. As this feeling, more than anything else, has given rise to the strong attachment to theories, which has always, and almost universally existed, and as a little reflection on its soundness may aid us in forming a correct estimate of the real value and importance of theories, it may be well to say a few words here upon the subject.

Sir Isaac Newton, it is well known, suggested that the phenomena of gravitation might possibly be explained and in some degree accounted for, by the presence and action, throughout all space, of a subtle and elastic ether. Professor Whewell remarks, also, that the presence of this pervading ether may remove a difficulty, which some persons find considerable, of imagining a body to exert force at a distance;[b] and I know, that this difficulty is often felt, even by minds of much strength and acuteness. It would seem that Newton him-[42]self was driven to the supposition of his ether, not merely as a convenient means of explaining the phenomena of gravitation, but as a necessary condition of these phenomena. He found it impossible to conceive of the existence of these phenomena, without the intervention of some material bond of connexion between the particles and masses of matter acting upon each other. In a letter to Dr. Bentley, he expresses himself in the following words: — "It is inconceivable that inanimate brute matter should, without the mediation of something of something else which is not material, operate upon, and affect other matter, without mutual contact, as it must do, if gravitation, in the sense of Epicurus, be essential and inherent in it. And this is one reason why I desired that you would not ascribe innate gravity to me. That gravity should be innate, inherent, and essential to matter, so that one body may act on another, through a vacuum, without the mediation of anything else, by and through which their action and force may be conveyed from one to another, is to me so great an absurdity, that I believe no man who has, in philosophical matters, a competent faculty of thinking, can ever fall into it."[c] But I do not see how the existence of such an ether can render any more intelligible the fact of gravitation, than it now is, without the ether. The everlasting and unanswerable WHY? [43] and HOW? Are not gotten rid of by this assumption, or by the discovery of a new phenomenon. We have only carried them a step farther from us, or brought them, apparently, a step nearer to us. In the first instance, the question was this; — *how*, or *why*, do all the solid particles of matter strive to approach each other, there being nothing but void

[b] Phil. Ind. Sci. vol ii. p. 210.

[c] Stewart's Philosophy of the Human Mind.

space between them? The question, in the second instance, becomes merely this; — *how*, or *why*, does this rare, ethereal medium, impalpable, imponderable, invisible, almost inappreciable by the most refined means of observation, draw together these solid particles? How, and by what mysterious and incomprehensible agency, does it hold the ultimate elements of matter in their relative positions, drag the avalanche from its rocky basis, call back the comet from its remote wanderings, and retain the planets in their orbits? I cannot see that one question is any easier of solution than the other. I cannot see that there is anything especially difficult, or unphilosophical, in the supposition, that gravitation consists exclusively and entirely in the tendency of the solid particles of matter to approach each other. Why may not this tendency exist as well without any intervening agent as with one? Furthermore, is it not true, that all our knowledge of the properties of matter leads to the probable conclusion, that its ultimate elements do not absolutely touch each other; that each on is surrounded by an [44] atmosphere, or space? Mossotti[7] and others have filled up these spaces with the assumed matter of electricity, or with the Newtonian ether; but this does not alter the essential constitution of matter, so far as this particular circumstance of the contact of its particles is concerned; for this supposed ether, however rare and attenuated it may be, must, after all, be composed of elements, or particles, as truly as matter itself. It is, at least, as difficult for us to conceive of any other constitution for the ether, as for solid matter. It seems probable, then, that there is no such thing in nature as absolute contact; at any rate, there is nothing unphilosophical in this conception of the ultimate arrangement of the elements of matter. Now if the supposed ultimate elements of common matter, or those of the assumed ethereal medium, can act upon each other through absolutely void spaces of infinite minuteness, there is no reason why the same elements may not also act upon each other through void spaces of infinite extent. I cannot see that the tendency of all bodies to approach each other, constituting the principle of gravity, is at all more incomprehensible and mysterious, than any other ascertained relation of the particles or masses of matter; and if it were so, I do not see how the intervention of the supposed matter of electricity, or of the supposed ether of Newton, can in any way aid, either in removing or diminishing the difficulty.

I think that a similar study of the theories of [45] light, electricity, chemical combination, and so on, would lead us to much the same conclusion. The theory, or hypothesis, in these, and in all analogous instances, might seem at first sight, to furnish material aid to the mind in its attempts to conceive and to comprehend the phenomena and relations to which the theory is applied. But we shall find, I think, that we get rid of one difficulty, so far as we do get rid of it, only by the substitution of another, no less formidable, in its place. The expedient is just about as successful in the accomplishment of its professed object, as that of the Indian philosopher, who placed the world on the back of a turtle; and it comes to much the same thing. The world would be well enough disposed of, if there were any stable resting place for the turtle to stand upon; and so our theories might indeed render more intelligible the subjects to which they refer, were not the theories, themselves, quite as difficult to comprehend as the phenomena and relations, which they profess to interpret and explain.

I shall conclude this chapter with a few remarks on the value and importance, in physical science, of theories, or hypotheses. Without qualifying, in any degree, the doctrine which I have been endeavoring to elucidate, that all science is independent of hypothesis, I am quite willing to admit, that the hypothesis has often been of service to science, in suggesting, guiding and directing its researches. I am willing to go further than this, as has already [46] been intimated, and to admit, at least the possibility, in some instances, that the researches thus suggested and directed, may lead, ultimately, to the positive demonstration of the assumed phenomena, constituting the theory. I am willing to admit with Professor Whewell, (the speculative tendencies of whose mind are very evident in all his writings,) the great difficulty, perhaps the impossibility, in many cases, of forming any definite conception of phenomena, or of reasoning upon them, without resorting to some hypothetical machinery, for the purpose of expressing, or interpreting their nature and relations.[d] But after all, I cannot avoid repeating the conviction, that an undue importance, and a false position, is still very generally assigned to these interpretations. The old and illegitimate usurpation of power, by the IDEAL PHILOSOPHY, in the empire of science, is even yet only partially destroyed; and the reign of EXPERIENCE, with that divine right, and absolute dominion, which constitute her inalienable prerogatives, has been only partially established. It is important to observe, farther, that the aids and uses, which may really be derived from hypotheses, will be in no way diminished, but increased rather, by assigning to them the subordinate character and station, which they ought always to occupy. If this is done, while their ability to advance the progress of science will not be in any degree [47] lessened, their mischievous tendencies in obscuring its perceptions, and in leading it astray, will be neutralized.

The influence of this particular element of false philosophy has been so disastrous in its effects on the progress of medical science, that I am especially anxious to exhibit it in its true light; and since the opinions which I have expressed may seem to be somewhat at variance with those which have been advocated by two most profound and elegant writers on the philosophy of science, — Sir John Herschel and Professor Whewell, — I will call to the support of the cause, which I have endeavored to vindicate, two other witnesses, certainly of not inferior competency and authority.

Sir Isaac Newton, as has already been stated, had his theory of light; and it is but reasonable to suppose, that whatever value it really possessed, must have been fully obvious to his own mind. At any rate, if we may judge from the nature and tendencies of the human mind, or from the history of science, he could not have been disposed, as its author, in any degree to undervalue its importance. Now, it is beautiful to witness with what true appreciation Newton regarded this his own theory — as well as those of others — of the properties and phenomena of light, which he had newly discovered; and with what lofty indifference, and disdain, almost, he cast it behind him. Amongst the intellectual elements, which contributed to his superiority, and which enabled [48] him to achieve a greatness and renown in

---

[d] Phil. Ind. Sci. vol. ii. p. 268-9.

the realms of science, now unrivalled and supreme, this rare quality was one of the earliest in its development, and most powerful in its operation. If he bowed at any time, or in any degree, his strong neck to the yoke of hypothesis, it was always with a perfect consciousness of his ability at will to shake it off, as the lion shakes the dew drop from his mane. It is well known, that his great discovery of the heterogeneous or compound nature of light was made in his early youth; and in his modest, manly, unassuming letter to Mr. Oldenburg, announcing his discovery, he says, in connexion with the subject before us: — "But to determine more absolutely what light is, after what manner refracted, and by what modes or actions it produces in our own minds the phantasms of colors, is not so easy. *And I shall not mingle conjectures with certainties.*"[e] In the discussions which followed the announcement of Newton's discovery, he had frequent occasion to refer to this matter, — of the value and importance of hypothetical explanations; — and I know, that in no other way can I do so much for the cause of sound philosophy, and for the gratification of its genuine lovers and disciples, as by quoting his golden words. In a reply to some rather captious animadversions of Father Pardies, he says: — "For the best and safest method of philosophizing seems to be, first [49] to inquire diligently into the properties of things, and establishing those properties by experiments, and then to proceed more slowly to hypotheses for the explanation of them. *For hypotheses should be subservient only in explaining the properties of things, but not assumed in determining them;* unless so far as they may furnish experiments. For if the possibility of hypotheses is to be the test of the truth and reality of things, I see not how certainty can be obtained in any science; since numerous hypotheses may be devised, which shall seem to overcome new difficulties." And again he says: — "Give me leave, sir, to insinuate that I cannot think it effectual for determining truth, to examine the several ways by which phenomena may be explained, unless where there can be a perfect enumeration of all those ways."[f]

Sir Humphrey Davy,[8] during the early period of his scientific researches, yielding to the impulses of a vivid and fertile imagination, suffered his mind to run riot in the creation of hypotheses. But in the full maturity and development of his powers, when his mind had become disciplined by habits of positive investigation and rigorous analysis, he abjured altogether this spurious philosophy of his youth; and no man ever saw more clearly and distinctly than he did the true character, and relations to science, of these hypothetical fancies. Amongst the many allusions to this subject, con-[50]tained in his writings, it is sufficient for my present purpose to cite only the following: — "When I consider the variety of theories that may be formed on the slender foundation of one or two facts, I am convinced that it is the business of the true philosopher to avoid them altogether. It is more laborious to accumulate facts than to reason concerning them; but one good experiment is of more value than the ingenuity of a brain like Newton's.[g] . . . "The theorizing habit in a sound mind can counteract only for a short time the love of seeing things in

---

[e] Phil. Trans. Anno, 1672.
[f] Phil. Trans. Anno, 1672.
[g] Life of Sir H. Davy. By Dr. Davy. Vol. i. p. 81, 82.

their real light; and the illusions of the imagination, in proportion as they often occur and are destroyed by facts, will become less vivid, and less capable of permanently misleading the mind."[h] . . . "Hypothesis should be considered merely as an intellectual instrument of discovery, which at any time may be relinquished for a better instrument. It should never be spoken of as a truth; its highest praise is verisimility; knowledge can only be acquired by the senses; nature has no archetype in the human imagination; her empire is given only to industry and action, guided and governed by experience."[i] . . . "I trust that our philosophers will attach no importance to hypotheses, except as leading to the research after facts, so as to be able to discard or adopt them at pleasure; treating them rather as parts of the scaffolding of the [51] building of science, than as belonging either to its foundations, materials, or ornaments."[j] I am entirely content with the position and importance thus assigned to theories and hypotheses by Newton and Davy.

CHAPTER V.

PROPOSITION SIXTH

ALL CLASSIFICATION, OR ARRANGEMENT, DEPENDS UPON, AND CONSISTS IN, THE IDENTITY, OR SIMILARITY, AMONGST THEMSELVES OF CERTAIN GROUPS OF PHENOMENA, OR RELATIONSHIPS; AND THEIR DISSIMILARITY TO OTHER GROUPS OF PHENOMENA, OR RELATIONSHIPS. ALL CLASSIFICATIONS, OR ARRANGEMENTS ARE NATURAL AND PERFECT JUST IN PROPORTION TO THE NUMBER, THE IMPORTANCE, AND THE DEGREE OF THESE SIMILARITIES AND DISSIMILARITIES.

Arrangement and classification of phenomena and relationships. Principles and grounds of this arrangement. Illustrations. Marble.

I HAVE said a good deal, in the course of this essay, about the *classification* and *arrangement* of phenomena and their relationships. I have said, again and again, not that science consists in phenomena and relationships, merely; but in [52] these phenomena and their relationships, *classified* and *arranged*. It is not enough to constitute science, that its materials should be discovered and ascertained; they must be brought together; they must be compared with each other; they must be analyzed, divided into groups, or families, placed in their appropriate positions; — in short,

[h] Ibid. p. 216.
[i] Ibid. p. 128.
[j] Life of Sir H. Davy. By Dr. Davy. Vol. ii. p. 128.

they must be *classified* and *arranged*. Until this is done, the materials themselves, heterogeneous, and jumbled together in disorder and confusion, are comparatively worthless; they have neither value nor significance. It is, indeed, by this process of classification and arrangement, that science is constructed. The phenomena and their relationships constitute the materials of the temple; it is by their classification and arrangement, only, that the temple itself is built up. The principles which are to guide us in this process, and the conditions of the process itself are, I think, clearly and succinctly stated in the proposition at the head of this chapter.[1]

I have endeavored, in some of the preceding pages of this essay, to illustrate some of its doctrines by an examination of the nature and the sources of our knowledge of marble; let us now endeavor to see by what process, and according to what rules, the elements of this knowledge, — the ascertained phenomena and relationships of marble, — are so classified and arranged as to convert them into science. One of these relationships is that of its particles to each other and to those of [53] all other material substances, through space. The property, constituting this relationship, is possessed by marble in common with all other material substances. The relationship, so far as these substances are concerned, is universal: in the possession of this property all these substances are alike; they are absolutely identical with each other; and by means of this identity they are constituted a class. They are called ponderable bodies: widely as they may differ in other respects, they all agree in this, that they are equally subject to the laws of gravitation. Now, there is another class of substances, or agencies, which do not possess this particular property; they are not subject to the laws of gravitation; this particular relationship does not touch them. Their freedom from this relationship, the absence of this single but fundamental property, constitutes them *a class*, entirely unlike that to which marble belongs; — they are called imponderable bodies. The classification, in the instance before us, depends upon and grows out of the presence, or the absence, of this single property; but the property itself is so important and so fundamental, that the classification itself assumes the same character.

In the second place, our knowledge of marble teaches us that it is a compound body — that it is formed by the intimate combination of several distinct substances. This property it possesses in common with all or nearly all other material substances as they exist about us. They may differ [54] indefinitely in all other respects, but they agree in this, that they are formed by the union of other substances; and the possession of this common characteristic constitutes them *a class*;— they are called compound bodies. It is found further, that other bodies, or substances, at least in the actual state of our knowledge, are not formed by the union of distinct elements; that they cannot be separated into other substances; they differ very widely amongst themselves in all other respects, but they are identical in this. Their entire and perfect similarity in this constitutes them *a class*; — they are called simple, or elementary bodies.

Again, it is found, that marble consists of carbonic acid in combination with another substance; and in this particular circumstance it resembles many other compounds; they constitute a class, and are call *carbonates*.

It is unnecessary to carry this illustration any further. I am not writing a treatise on the physical sciences; and I only wish to present, in as few words as possible, the principle on which all classification in these sciences must rest; and to show in what it consists. [55]

PART SECOND

THE

PHILOSOPHY OF MEDICAL SCIENCE [57]

"Ars medica tota observationibus." *Frederick Hoffman.*

"Hypotheses and imaginary suppositions never should be admitted either into philosophy, or the medical science." *William Hillary.*

"From what we have said before, it appears, that all the knowledge that we have of the virtues, operations, and effects which all plants, drugs, and all medicines that we yet know, have in and upon the human body, has been obtained by observation and experience; neither does the human mind seem capable of acquiring that knowledge by any other means." *Ib.*

"It is not from ingenious reasoning, or fine-spun theories, that we should estimate the value of a remedy, but from the effects actually produced by it in a majority of cases." *Nathan Smith.*

"La medecine ne s'enrichit que par les faits." *Broussais.*

"*Calcul des resultats,* seule manière infaillible d'apprécier la valeur des methodes en medecine." *Dupuytren, Mem. de l'Acad. Roy. de Med.* 1824.

"It is only by computation, founded upon large averages, that truth can be ascertained, and hence the danger of founding a general practice on the experience of a single case, or a few cases." *Sir Gilbert Blane.*

"The materials of just pathology can be drawn only from large masses of observation assembled and arranged in the order of their subjects; nor can durable improvements in practice be established on less than full and luminous evidence." *Edward Percival.*

"Through medical statistics lies the most secure path into the philosophy of medicine." *Henry Holland.*

"Sufficit si *quid* fit intelligamus, etsi *quomodo* quidque fiat ignoremus." *Cicero.*

"It appears to me that the physician, who ascertains half a dozen of important facts, performs a more valuable though a less splendid achievement, than he who invents a dazzling theory." *Samuel Black.*

"Analogy, the fruitful parent of fallacious conclusions." *William Woolcombe.*

"Analogy, that fertile source of error. *Liebig.* [58]

PART SECOND

MEDICAL SCIENCE

PRIMARY PROPOSITIONS

Proposition First. All medical science consists in ascertained facts, or phenomena, or events; with their relations to other facts, or phenomena, or events; the whole classified, and arranged.

Proposition Second. Each separate class of facts, phenomena, and events, with their relationships, constituting, as far as they go, medical science, can be ascertained in only one way; and that is by observation, or experience. They cannot be deduced, or inferred, from any other class of facts, phenomena, events, or relationships, by any process of induction, or reasoning, independent of observation.

Proposition Third. An absolute law, or principle, of medical science consists in an absolute and rigorous generalization of some of the facts, phenomena, events, or relationships, by the sum of which the science is constituted. The actual, ascertainable laws, or principles, of medical science are, for the most part, not absolute but approximative.

Proposition Fourth. Medical doctrines, as they are called, are, in most instances, hypothetical explanations, or interpretations, merely, of the ascertained phenomena, and their relationships, of medical science. These explanations consist of certain other assumed and unascertained phenomena and [59] relationships. They do not constitute a legitimate element of medical science. All medical science is absolutely independent of these explanations.

Proposition Fifth. Diseases, like all other objects of natural history, are susceptible of classification and arrangement. This classification and arrangement will be natural and perfect just in proportion to the number, the importance, and the degree of the similarities and the dissimilarities between the diseases themselves. [60]

PART SECOND

THE PHILOSOPHY OF MEDICAL SCIENCE

CHAPTER I.

Definitions. Anatomy; Topographical; General; Microscopic; Chemical; Comparative. Physiology. Etiology. Therapeutics.

MEDICAL science, in the comprehensive meaning here attached to it, includes the whole science of organization, or life. This science is made up of many integral constituents; it consists of a considerable number of distinct and separate classes of phenomena and relationships, constituting so many individual branches, or departments, of the entire science; and before proceeding to the principal subject of my essay, it is necessary, briefly and distinctly, to define these branches — to state what the phenomena and relations are, in which each and all of them consist.

The first of these departments is that which relates to the material structure, or organization, of living beings. This department is called anatomy. Inasmuch as it involves a knowledge of all the conditions which combine to constitute the [61] structure and conformation of living bodies; and inasmuch as these conditions are many and diverse, this primary division naturally separates itself into several sub-divisions, founded upon this number and diversity of conditions, which unite to constitute the structure.

The first of these sub-divisions relates to the manifest and sensible properties and relations of each separate individual part, or organ, of a living being. It involves a knowledge of the size, form, color, consistence, specific gravity, position, and arrangement, of each and all of these single parts, or organs. It is called special, or topographical anatomy. Every separate individual, in the two great organic kingdoms — of animal and of vegetable life — has its own peculiar and characteristic topographical anatomy, or conformation; constituting, so far as structure is concerned, its individual peculiarity. This may be called individual topographical anatomy. Each sex also, where the sexes are separate, both in the animal and the vegetable world, has its peculiar structure; and this peculiarity of structure constitutes the anatomy of the sexes. The obvious structure of the several parts and organs of living beings differs, more or less, during the successive periods of their growth and decay; and these differences constitute the topographical anatomy of the several ages, or periods, of life.

The second division of anatomy relates to the obvious structure and properties, not of the indi-[62]vidual parts, or organs, which make up a living body, but of the organic elements, or tissues, as they are called, of which these single parts are composed. This is called general, or physiological anatomy. It is found, by observation, that every separate part, or organ, is not simple and homogeneous, but complex and heterogeneous, in its anatomical composition; and general or physiological anatomy consists in a knowledge of all the sensible properties and relations of these separate organic elements, or tissues, in whatever part or organ of the body they may be found. While topographical anatomy, for instance, informs us of the size, shape, arrangement, position, and relations of the heart, considered as a whole, general anatomy examines its organic composition, and teaches us the properties of the several elements, or tissues, — the muscular, the serous, the cellular, and so on, — which unite to constitute it what it is.

The third division of anatomy relates to the more hidden and delicate structure of organized bodies. With the aid of lenses, it pushes its investigations far beyond the line which limits the unassisted senses, and strains its vision to detect the ultimate and final arrangement of the primordial organic elements. It traces the capillary vessels to their minutest anastomoses; it unravels the smallest muscular bundles to their ultimate fibres; it follows the gossamer thread of the nerve to its final termination; it measures the diameter [63] of the blood-globule, and estimates the thickness of its colored envelope. It is called minute, or microscopic anatomy.

The fourth division of this department relates to the chemical composition of the organs, tissues, and fluids of living bodies. All these organs, tissues, and fluids are made up of the elements of common matter; and this division teaches us what these elements are, and in what proportions they are combined. It may be called chemical anatomy. The structure of the human body, in its several sub-divisions, constitutes human anatomy; that of vegetables, what may be called vegetable comparative anatomy; and that of animals, below man, animal comparative anatomy.

The second great department of the science of life is that, which relates to the actions or processes, which result from, or are connected with, the structure or organization of living beings. This department of the science is called physiology. Its sub-divisions, or branches, correspond very nearly to those of anatomy. So far as observation enables us to judge, every peculiarity and variety of structure is associated with a peculiarity and variety of action. Each organ, each apparatus of organs, each elementary tissue, plays its own part, performs its own specific duty, accomplishes its peculiar and individual office, in the living economy. The actions and processes, which take place in, and are effected by, each part, or organ, or tissue, constitute the physiology of this part, or [64] organ, or tissue. There is, therefore, a physiology of each organ, and of each elementary tissue; there is a physiology of the sexes, and of each successive period of life; there is a comparative vegetable physiology, a comparative animal physiology, and so on. Anatomy consists in the entire structure of organized bodies; physiology consists in the natural and regular actions and processes connected with this structure; life is the aggregate sum of the two, — of the structure and its functions. The entire science of life, in its natural or normal condition, is contained in these phenomena, of structure and function, and in their various relations. The entire natural history of living beings, in all their infinite variety of form, structure, and function, — from the hyssop upon the wall, to the cedar of Lebanon, and from the microscopic animal monad, to man, — is contained in these phenomena and their relationships, classified and arranged.

But inasmuch as the structure of living beings, and the actions and processes connected with this structure, are subject to various derangements, and departures from their natural and normal condition, we have a third fundamental department of the science of life, consisting in the phenomena, and their relations, of this altered structure, and of these disordered actions and processes. This department is called pathology. It is coextensive with the two preceding departments. Alterations of the structure, appreciable in any way by the [65] senses,[1] constitute what has been called morbid or pathological anatomy; derangements in the actions and processes, connected with the structure, constitute what has been called, simply, pathology. The former may be more properly termed structural or organic pathology; and the latter, functional pathology. The entire science of pathology, or disease, consists in these phenomena and their relationships, classified and arranged. These relationships are of a threefold character. The first are those which exist amongst the phenomena themselves; which relations, with the phenomena, constitute, as has just been said, the department itself of pathology. The second are those which exist between these phenomena, on the one hand, and all those substances, agencies and influences, of whatever sort or character, which occasion or give rise to the phenomena, — which precede, and stand to them, as causes. These relationships constitute the sub-division of pathology, which is called etiology, or the science of the causes of disease. The third class of relations are those which exist between the phenomena of altered structure and disordered function, on the one hand, and all those substances, agencies and influences, on the other, the properties and operation of which are to arrest the progress of these phenomena, to restrain them within such limits as are compatible with life, to shorten their duration, to modify them in one way or another, or to remove them alto-[66]gether, — thus restoring the structure and function from a pathological to a physiological condition — from disease to health. These relations constitute that sub-division of pathology which is called therapeutics.[2]

CHAPTER II.

PROPOSITION FIRST

ALL MEDICAL SCIENCE CONSISTS IN ASCERTAINED FACTS, OR PHENOMENA, OR EVENTS; WITH THEIR RELATIONS TO OTHER FACTS,

OR PHENOMENA, OR EVENTS; THE WHOLE CLASSIFIED AND ARRANGED.

General prevalence of false notions. Medical science consists, exclusively, in the phenomena and relationships of life, classified and arranged. Anatomy. Physiology. Illustrations. Germination of seeds. Conditions of germination. Phenomena of germination. Respiration; its phenomena.

IF it is true, even in physical science, as I have endeavored to show, that the doctrine stated in my first proposition, is only partially and imperfectly recognized, it is so to a much greater extent in regard to the same doctrine in its application to medical science; and the remarks already made upon this doctrine may be repeated with still less qualification, and with more emphatic significance, in connexion with my present subject. The fundamental and primary truth, that all medi-[67]cal science consists in the appreciable phenomena of life, with their relationships, classified and arranged, and in nothing else, has never been generally admitted and received. This science, to a vastly greater extent than any other, has always suffered, and still continues to suffer, from the general prevalence of a spurious philosophy, and from vicious or imperfect methods of investigation; and one element in this false philosophy, leading to these mistaken methods, is to be found in the inadequate conception or half-belief of the doctrine above stated. Here, as in physical science, with very few exceptions, men, claiming to be disciples of the Baconian philosophy, eloquent in their praises of what they call *inductive reasoning*, and full of earnest declamation against the dangers and the prevalence of false or premature generalizations, and of hypothetical speculation, have failed to see more than half the truth, and have, oftener than otherwise, fallen headlong into the errors which they were so ready to condemn. The feeling has been much more common in medical, than in physical science, that although facts and their relations might, indeed, and must constitute the *foundation* of the science, the science still consisted in something more that these facts and relations; — that *upon* these latter the science itself was to be somehow built up, by that magical and creative process of the mind, — that evil genius of medical science, — called, indeed, *induction*, but differing, when stripped of its dis-[68]guises, in no single function or attribute, from that speculation, the place of which it professed, with promises as loud and pompous, as they have proved to be barren and empty, to occupy. The feeling has been, and still is, — as much, almost, since the time of Bacon as before, — that the science is in the *inductive or reasoning process, superadded to the facts and their relations*, more than in these latter themselves. Here, at the commencement of this part of my essay, I wish to enter my protest against this doctrine, in all its forms and modifications. I wish to show, that the science of medicine consists *in the phenomena of life, with their relationships, classified and arranged*, — WHOLLY, ENTIRELY, ABSOLUTELY. I wish to show, that these elements constitute, — not the foundation upon which, nor the materials, merely, with which, the science is to be subsequently constructed, by some recondite and logical process of the reason, — but that they *are* the science, and the whole science, already constructed, and so far completed; and that nothing can be superadded to them, by any act of the mind, which can in any way increase their value, or change their character.[1]

This doctrine, in its relation to anatomy, needs but little, if any, illustration. It is so obviously true here, that there is hardly any room for misconception or doubt. Medical science, so far as anatomy is concerned, consists so manifestly in the physical phenomena connected with organization, and in no-[69]thing else, that there is no necessity for any formal discussion of the subject; and I will, therefore, pass at once to the consideration of the doctrine before us in its connexion with the other and more complicated branches of the science.

It has already been stated, that the physiology of any living being consists in the sum or aggregate of its normal actions, and of their relationships. Now in order to see whether these actions and relations, classified and arranged, do, or do not, constitute the whole of the science, let us examine some of the processes and series of processes, which are carried on in the organic structure, both of vegetable and of animal life. Let us look first at the germination of a seed, and see what the actions and relations are in which it consists. We find, in the first place, that there are three indispensable or essential conditions, if either one of which be wanting, the changes or actions in the seed, the sum of which constitutes germination, will not take place. The first of these conditions is that of temperature, or the degree of heat, in the midst of which the seed is placed. The temperature necessary to the process of germination varies somewhat with different kinds of seeds; but its range does not extend much below the freezing point, nor above 100° of Fahrenheit. Without these limits the susceptibility of the seed to take on, and to go through with, the processes constituting germination, is not awakened into action, and it may be wholly destroyed. The second [70] condition consists in the presence of water. If this latter substance is entirely wanting, the processes do not take place. The third condition consists in the presence of atmospheric air. If either of these conditions is wanting, the seed does not germinate; it remains quiescent, or its peculiar structure and susceptibilities are destroyed. When, however, the foregoing relations are established, the process of germination is set up and carried on, and this process consists in the following changes. Certain parts, or organs, of the seed, which are called its *cotyledons*, are increased in size; and a part at least of this increase is occasioned by the reception of the surrounding water into the minute cells of their structure. This swelling of the cotyledons, ruptures their external investing membrane; they separate somewhat from each other, and thus their original relative position is changed. The consistence of the cotyledons is also considerably diminished. Cotemporaneous with these changes in the cotyledon, another part or organ of the seed, termed the *plumula*, is enlarged, and extends itself in an upward direction; while still another, termed the *radicle*, is also enlarged, and extends itself in an opposite direction. Accompanying these obvious changes in the volume, the consistence, and the relative positions of the several organs, or anatomical parts of the seed, there are others which have taken place in its chemical composition. The proportion of ultimate elements, — carbon, [71] oxygen, and hydrogen, — originally constituting this composition, is found to be altered, a portion of the carbon having disappeared from the seed, and united with a portion of the surrounding oxygen of the atmospheric air. The insipid and farinaceous substance of the cotyledons has become sweet and mucilaginous; their albuminous and amylaceous parts having been converted into gum and sugar. These changes are accompanied by a considerable elevation in the temperature of the seed; and all the processes are, furthermore, to a certain extent, influenced by the degree of light that is present.

Such, very briefly stated, are the phenomena and relationships constituting the physiology of germination. When these phenomena and relationships have been fully and positively ascertained, and classified, the science of physiology, so far as germination is concerned, is complete. These facts and relations are not to be used as materials, merely, wherewith the science is subsequently to be created by some process of reasoning. They *are*, already, the science, and the whole science. There may be difficulties, — many and great, — in arriving at a full and absolute knowledge of all the processes that have been spoken of; there may be difficulties in ascertaining positively all their relationships, — in referring each to its proper mechanical, chemical, or vital cause, or to various combinations of these several agencies; there may be much ingenious speculation about [72] these processes and relationship; but, after all, the physiology of germination will be found to consist, solely and exclusively, in these ascertained processes and relations. No act of the mind can add anything to what has already been done. These phenomena and relationships *are not to be converted into* the science of physiology, by an inductive process; — they *are* the science; — the science consists of these and of nothing else.

The several obvious acts and changes constituting the function of respiration, in the higher classes of animals, are very well ascertained. By a strictly mechanical process, the atmospheric air is introduced into the lungs, and after a short continuance there, again driven out. The expelled air is found, on examination, to have parted with a portion of its oxygen, during its presence in the lungs, and to have acquired an undue proportion or quantity of carbonic acid. On farther examination it is found, that the carbon in the acid, or the acid already formed, has been derived from the venous blood; and that the oxygen, which has disappeared from the respired air, has been either absorbed by the blood, or united with the carbon to form the acid. Cotemporaneous with these changes, the venous blood has been altered in its color, and in some other of its properties, by which it has been converted into what is called arterial blood. And here, as in the germination of a seed, the whole science of physiology, so far as respiration is concerned, consists [73] in the phenomena and their relationships. When the phenomena and relationships, constituting this function, have been ascertained and classified, throughout the entire range of living beings, the physiology of respiration is completed. No reasoning upon these phenomena, no speculations about them, can give them any new character, or make them any more legitimate elements of science than they already are. And a similar study of each and of all the functions of living beings will lead us to the same results. But my purpose here is only to establish and illustrate a doctrine, not to teach physiology; and such a study would be an unnecessary waste of labor and time.

This doctrine is just as true in its application to pathology, etiology, and therapeutics, as to physiology. In each of these fundamental branches of medical science, the science consists in the phenomena and relationships, with which the particular branch is concerned, classified and arranged, and not in any superadded reasonings or inductions of our own. But inasmuch as the doctrine has already been somewhat fully developed, and as it will receive other incidental illustrations in the further prosecution of my subject, I will say no more of it here. [74]

CHAPTER III.

PROPOSITION SECOND

EACH SEPARATE CLASS OF FACTS, PHENOMENA, AND EVENTS, WITH THEIR RELATIONSHIPS, CONSTITUTING, AS FAR AS THEY GO, MEDICAL SCIENCE, CAN BE ASCERTAINED IN ONLY ONE WAY; AND THAT IS BY OBSERVATION, OR EXPERIENCE. THEY CANNOT BE DEDUCED, OR INFERRED, FROM ANY OTHER CLASS OF FACTS, PHENOMENA, EVENTS, OR RELATIONSHIPS, BY ANY PROCESS OF INDUCTION OR REASONING, INDEPENDENT OF OBSERVATION.

Extent of erroneous notions.

THE development of the doctrine, enunciated in the above proposition, will constitute a very prominent portion of my essay. We are approaching, I think, one of the strong holds of error in the philosophy of medical science, and a good part of my remarks thus far have been preliminary, only, to the inquisition which we are now prepared to institute into the nature and extent of this error. I have already said, that even in physical science, the doctrine, that each separate class of phenomena and relationships can be ascertained only by direct observation of these phenomena and relationships, themselves; and that a knowledge of one class cannot be deduced or inferred from the knowledge of any other class, by any process of [75] the pure reason, is only partially admitted. But it is in medical science, especially, that this great and fundamental principle has been most generally and extensively disregarded. The feeling has been almost universal, and it still continues so, that the several classes of phenomena and relationships, constituting the science are somehow so allied to each other, that a knowledge of one class may be, to a greater or less extent, deduced from a knowledge of the other classes. The prevalent idea is, that this connexion between the different branches of medical science is of such a character, that a knowledge of one branch may lead, by some deductive process, as it is called, to a knowledge of other branches. We are constantly told, for instance, that physiology is founded upon anatomy; that pathology is founded upon physiology; that therapeutics is to be *deduced*[1] from pathology, and so on. This assumed connexion between these and between other branches of medical science; this alleged dependence of a knowledge of one series or class of phenomena and relationships upon a knowledge of another class, or series, constitutes the principal ingredient in the error to which I have alluded, and which I wish to expose and remove; and the opposite doctrine of the entire dependence of all our knowledge of each series of phenomena and relationships upon direct observation of each particular series, is that which I wish to vindicate, and set up in its place. [76]

CHAPTER IV.

Our knowledge of anatomy not dependent upon our knowledge of other branches of medical science. Our knowledge of one branch of anatomy does not include the knowledge of any other branch.

THAT our knowledge of the structure and composition of all living bodies is the exclusive result of observation is so plain and obvious a truth as to stand in need of no illustration; and we find, accordingly, that the influence of a false philosophy, and a vicious method of investigation has been less felt in this than in any other department of the science of life. I shall, therefore, have but little to say in this chapter. I wish to remark, however, that our knowledge of each sub-division of this special department is wholly independent of our knowledge of other sub-divisions. One kind of structure, or composition, is not to be deduced from another. Our acquaintance with each sub-division is the exclusive result of our examination of that particular sub-division. Topographical anatomy is to be learned only by direct study of the form, the volume, the color, the consistence, the position, and so on, of the several individual parts or organs. Physiological or general anatomy, in its turn, is to be learned only by studying the properties of the several elementary [77] tissues which go to make up the organs; it cannot be deduced, or inferred, from the former. A knowledge of the anatomy of any one sex does not involve a knowledge of the anatomy of its corresponding, opposite sex. The intimate and minute structure of the several organs and tissues can never be inferred from their obvious physical qualities. For our knowledge of this we must rely wholly upon minute and microscopic examination. And the same thing is true of the chemical composition of the organs and tissues. This can in no way be inferred, or deduced, by any process of reasoning, from their other properties. No knowledge, however, accurate, of the conformation of the brain, or the liver; no knowledge, however, accurate, of the shape and arrangement of the ultimate anatomical elements of the two organs, could ever have furnished us with the remotest intimation of the chemical constitution of one or the other. The knowledge of this latter is to be obtained by direct observation, through the aid of chemistry, and in no other way. So, also, of vegetable anatomy in all its sub-divisions. Each of these latter is independent of the corresponding sub-divisions of animal anatomy; and each, also, is independent of the other, in its own department. Each must be learned by the direct study of its own characteristic phenomena: — no one can be inferred from either or from all of the others.[1] [78]

CHAPTER V.

Our knowledge of physiology not deducible from our knowledge of anatomy. Qualifications. Final causes. Illustrations. Brain. Stomach.

MY object, in the present chapter, is to show, that the actions of the organs and tissues which constitute living bodies can be ascertained only by direct observation of the actions themselves; — that they cannot be inferred from the structure of the organs and tissues; or, in other words, that physiology cannot be deduced from anatomy. Before proceeding, however, to do this, it is necessary to make one qualification, or explanation. This qualification grows out of, and depends upon, the great principle of the adaptation of means to ends. But this principle, as we call it, can hardly be regarded as an exception to the doctrine which I wish to set forth. It consists simply in the fact, always observed, when our means of observation are adequate, that throughout the universe, means are invariably and perfectly adapted to ends; and the qualification to which I allude consists merely in the application of this universal fact to the subject before us.[a] [79] Thus, when we examine the structure and conformation of the skull, we might justly and safely come to the conclusion, that it is intended to contain some organ or substance, which requires to be protected from the mechanical action of external bodies. In conformity to this great law of adaptation, running through all nature, and ascertained by observation, we might say, that the structure of the skull presupposes this as one of its functions. In conformity again to this law, and having ascertained by experience the action and offices of a valvular apparatus; form an accurate knowledge of the internal structure,

---

[a] This is the doctrine of *final causes*, — a doctrine, which, notwithstanding the objections of Geoffroy Saint-Hillaire, and some others, it seems to me utterly impossible not to see, written legibly and boldly, throughout all organized nature, — so legibly and boldly, that he who runs may read. The theological relations of this doctrine I in no way allude to on the present occasion. They have no bearing whatever on the question before us. The strength and soundness of the great argument of Paley and others, drawn from this universal fact of adaptation and apparent design, in favor of the existence and agency of an intelligent designer, has nothing to do with the fact itself. Whatever may be thought of the former, it seems impossible that the latter can be denied. The existence of what are called *final causes*, in physiology, or the fact itself of the adaptation of the organs to their uses, is an observed fact, just as obvious as any other fact or phenomenon, whatever, in nature.

and mechanical arrangement, of the heart, and its connexions with the venous and arterial tubes, we might justly and safely come to the conclusion, that its office, — so far as this structure and arrangement, and these connexions, are concerned, — is to receive from one set of tubes, and to [80] transmit to the other set, some kind of a circulating fluid. So, we might infer, in the same way, that the stomach, the gall bladder, the urinary bladder, the uterus, and so on, are intended to act as reservoirs; and that the several canals, leading to and from these reservoirs, as well as the other canals in the body, are intended for the passage or transmission of some kind or kinds of substances. From our knowledge of the properties and relations of light, and in accordance with the same principle of adaptation, we might conclude, that the transparent cornea is intended for the transmission of light, and that the crystalline lens has for its function the refraction of this same transmitted light, and the consequent formation of the images of visible objects upon the nervous expansion at the bottom of the eye. An examination of the structure of a bony articulation might fairly lead to the conclusion that its surfaces are intended to move upon each other. But in these, and in all analogous cases, we can go no further. Our inferences, or deductions, as we call them, are very limited in extent; and they consist only in the particular application of the law of *final causes*, or the general fact of the observed relation of means to ends. The conformation of the skull gives us no intimation of the character, the properties, or the uses of the substance contained within it. The arrangement of the heart, and its dependencies, throws no light upon the nature of the offices of the fluid which [81] they are designed to circulate. This fluid, for aught that the anatomical arrangement for the organs teaches us to the contrary, might be water, or milk, or air, as well as blood. The same thing is true of the other canals, and of the several reservoirs of the body. Their structure points out only their general, and not their particular and positive uses. No one could ever have inferred, from any *à priori* reasoning, that the gall bladder was intended as a receptacle for bile, or the urinary bladder as a reservoir for the urine.[1]

It is a matter of very little importance, what organs, or what functions, of a living body are selected, for the purpose of illustrating the doctrine of this chapter; any and all of these organs and functions will answer this purpose. No knowledge, however, complete, of the structure and composition of a seed, could ever have shed any light upon the vital actions which it is capable of manifesting; no acquaintance, however, perfect, with the anatomical and chemical elements of its plumula and radicle, could ever have furnished the remotest intimation of the tendency in one to stretch upwards, to form the stem, and in the other, to reach downwards, to form the root of the new plant. Is there anything in the obvious physical properties of the glands of the human body — is there anything in their chemical composition, or in their minute, molecular arrangement, — from which even the obscurest and most shadowy glimpse could have been obtained, of the several [82] offices which they are destined to perform? Could the scalpel of the dissector, or the lenses of the optician, or the retort of the analyst, or all combined, have ever revealed to us the power of the liver to secrete bile, of the kidneys to secrete urine, or the mammary glands to secrete milk? Let us suppose that our anatomical knowledge of the brain had reached its ultimate limit of accuracy and perfection — that its complicated and delicate meshes of tubes and fibres had all been unravelled — that its intricate connexions and dependencies had all been ascertained — that no element or condition of its material organization had escaped us; — would all this knowledge[2] have furnished us with any information as to the part which it plays, and the offices which it performs, in the living economy? Has human reason any power sufficiently subtle and acute, to have extorted from this structure the secrets of its vital capacities? Has she any wand of so potent magic, as to have opened the mysteries wrapped up in the organization of the brain? Could she have detected, even in the ultimate recesses of this organization, if she could have penetrated thither, the latent power of the will — the yet unawakened capacity of sensation — the slumbering, but manifold and stupendous energies of emotion and thought? Were not some of the noblest functions of the brain, indeed, actually attributed by the *à priori* physiologists of former times to other and remote portions of the body? [83] Is there, in short, any conceivable process of induction, by which the physiology of the brain could have been derived from its anatomy? Certainly, there can be but one answer to these questions.

It may possibly be said by some, that this illustration is not a fair one, on account of the very peculiar functions of the brain, partaking, as they do, of what is regarded as immaterial or spiritual, in its nature. I cannot see that there is any difference, in regard to the subject before us, between the brain and its functions, and any other organ or apparatus of the body, and its functions. What is true of one, will be found, I think, with the qualifications already made, to be true of all. Let us look, for a moment, at some one of the organs and functions, of a purely material character. How is it with the structure of the stomach, and its functions? Could the latter have been deduced, by any act of the reason, independent of observation, from the former, any more readily than the functions of the brain could have been deduced from the structure of that organ? Is there anything in the anatomical character — topographical, microscopic, or chemical — of the mucous membrane of the stomach, that includes, or presupposes, in any way, its peculiar vital properties? For aught that mere anatomy teaches to the contrary, the function of digestion might just as well have been carried on by any other portion of the mucous membrane, as by that of [84] the stomach.[3] In short, here, as everywhere else, each separate class of phenomena and relationships can be ascertained in only one way, and that is, by direct observation of the phenomena and relationships themselves. For our knowledge of the offices and uses of every tissue, of every organ, of every apparatus, in the body, we must depend exclusively upon observation of these particular offices and uses, themselves; in no case can we derive this knowledge from any other sources.[4]

CHAPTER VI.

Our knowledge of pathology not deducible from our knowledge of physiology. Qualifications. Illustrations. Inflammation. Differences in the susceptibility of different organs to this process. These differences not to be accounted for on physiological grounds. Gastritis. Other diseases.

IN the preceding chapter, I have endeavored to exhibit the independent nature of our knowledge of physiology. I propose, in the present, to treat, in the same manner, of pathology. I wish to show, in the first place, that our knowledge of the morbid processes and susceptibilities of the several organs and tissues of the body cannot be inferred or deduced from our knowledge of their healthy processes. *Pathology is not founded upon physiology. The latter is not the basis of the for-*[85]*mer. The one does not flow from the other. Our knowledge of the one does not presuppose our knowledge of the other.* These assertions are so directly opposed to the common doctrine upon this subject, that it becomes necessary to show their truth and soundness, by a somewhat full development and illustration. It will not do, here, to say with Rousseau,[1] "*Ma fonction est de dire la vérité, mais non pas de la faire croire.*"[a] On the contrary, my function is, not only to speak the truth, but also, and especially, to show that these doctrines *are* sound and true. In the discussion of this subject, I leave wholly out of consideration the question of the dependence of our knowledge of pathology upon our knowledge of anatomy. If the healthy actions, and the natural uses of a part, or organ, or apparatus, cannot be inferred from its anatomical composition, much more evident is it, that the same thing is true of its diseased actions.

The doctrine, thus stated, and which I now proceed to illustrate, is subject to certain apparent qualifications, which ought, in the first place, to be pointed out. I have already said, that every simple and direct relationship, is constant and invariable. Supposing now the physiological actions and relationships of the body to be fully as-[86]certained, we may safely conclude, independent of positive experience, that a change in these relationships will be followed by a change in the actions themselves, and in the results of these actions.[2] Thus, after physiology has taught us, as far as it can teach us, the action of the oxygen of the atmosphere upon the blood, we may safely and positively conclude, prior to all experience, and independent of it, that if this action is interrupted, all the subsequent physiological processes with which it is connected will be also, and necessarily, more or less disturbed; and the same thing is true, of all physiological actions and relationships.

Again, inasmuch as the integrity of the mechanical contrivances and apparatuses of the body is necessary to the perfect performance of their offices, and inasmuch as

---

[a] My function is to speak the truth, but not to make it believed.

these contrivances and apparatuses are manifestly liable to injury, from external and obvious mechanical causes, it follows that, as in the former case, we may, so far as these mechanical relationships are concerned, infer the effects and consequences of such injuries, independent of absolute experience. Independent of any knowledge derived from observation of the fact itself, we might be quite certain, that an injury, or the destruction, of the aortic valves, would be followed by more or less disturbance of the function of circulation; and that the fracture of the femur would impede or destroy the act of locomotion, so far as this bone is concerned in the [87] performance of this function. So, from the conformation of the skull — from the manifest *design* for the protection of its contents from mechanical injury, which this conformation exhibits — we might, independent of any other or further knowledge derived from experience, very safely and confidently conclude, that such mechanical injury of its contained organ, or organs, would be followed by serious disturbance of the functions of the latter. But even here, we could go no further; how this disturbance would manifest itself, and in what it would consist, it would be utterly impossible for us to say, or to conjecture. Our knowledge of the functions of the brain would not enable us to predict, independent of actual experience, what particular manifestation of these functions would be injured or destroyed by any particular form of mechanical injury. No process of deduction, or of *à priori* reasoning, could lead us to the knowledge, that one species of injury would produce coma, another convulsions, and so on.[3]

In accordance with the same law of the invariableness of relationships, having ascertained, by observation, the forms and modes of diseased action, to which a certain part or tissue of the body is subject, we might infer, with a reasonable degree of certainty, that other parts or tissues, resembling the former in composition and in function, would be subject to similar forms and modes of diseased action. But, inasmuch as the re-[88]semblance in these cases is almost always one of a greater or less similarity, and not one of absolute identity, our *à priori* conclusions must be probable, only; not positive. The differences of structure and function, between analogous parts or tissues, though apparently slight, may still be sufficient to give rise to very great differences in the character and the importance of the lesions to which they are subject. Thus, notwithstanding the close similarity of structure and function, between the mucous membrane lining the trachea, and that lining the smaller bronchial ramifications, we find that the two are subject to important differences of morbid action, when they are attacked with acute inflammation; the former throwing out fibrine upon its surface, in the form of a membrane; the latter secreting only mucus. So, in acute inflammations of the serous covering of the lungs, and of the abdominal viscera; notwithstanding the near resemblance in the structure and functions of these two membranes, and notwithstanding the almost exact similarity in their appreciable pathology, the former is attended with a small degree of danger, the serum being absorbed, and adhesion taking place between the corresponding surfaces of the membrane; while the latter, at any rate after serum and fibrine have been thrown out, is almost invariable followed by a fatal termination.

Again, physiology having taught us the connexion between certain organs of the body, and [89] their dependencies and influences upon each other, we might properly enough conclude, that a similar connexion and dependence would show itself in their morbid actions and susceptibilities. Having ascertained, for instance, the existence of this connexion, and of these dependencies, between the several organs, in the female, constituting that extensive and complicated apparatus, for the continuance of the species, we might reasonably suppose, that a morbid condition of one portion of this apparatus would not be without influence upon the other portions. But, here, as in some of the cases already spoken of, our *à priori* conclusions could only be more or less probable; they could have nothing whatever of a certain and positive character. Actual observation would, in many instances, destroy instead of confirming them.

Finally, so intimate, and complicated, and manifold, are the physiological relationships of the living economy; so closely is each part connected with the rest; so readily and powerfully do these parts act and react upon each other; so complete, in many instances, is the union and the coöperation of the mechanical, the chemical, and the vital processes, that independent of actual experience, we might safely conclude, that an injury inflicted upon one part of the body might often affect, more or less seriously, many other parts; and that a disturbance, or suspension, of any one of the three great processes might, in many cases at any rate, disturb, or suspend, the other two. Having ascer-[90]tained, by observation, the complicated physiology of the circulation and of respiration; having ascertained the existence of various mechanical, chemical, and vital actions, and their necessary coöperation in order to produce a certain result, consisting in the circulation, the oxygenation, and the decarbonization of the blood; it would follow, as a matter of course, and independent of positive experience, that a disturbance, or suspension, of one of these associated actions, should disturb, or suspend, the others. Inasmuch as the integrity of the mechanical contrivances for the repeated exposure of the blood to the influence of the atmospheric oxygen is necessary, to secure this exposure, and inasmuch as this exposure is necessary, in order that the oxygenation, and decarbonization of the blood should be effected; and inasmuch as this change in the blood is essential in order to prepare it for answering its purposes in the vital processes of the body, it follows, necessarily, that any disturbance, or imperfection, in the first, mechanical process, will be followed by corresponding disturbances and imperfections in the subsequent and associated chemical and vital processes. In the same way, also, having ascertained that certain substances are eliminated from the body by the physiological actions of the liver, and the kidneys, we might justly come to the conclusion, without waiting for the positive teachings of experience, that the retention of these substances within the system would be followed by un-[91]favorable results. But, even in these cases, we could go no further. Although we might safely enough predict, that the non-oxygenation of the blood, and the failure of the liver, and the kidneys, to eliminate and to remove from the system their appropriate excrementitious and effete secretions, would be followed by unfriendly and probably fatal consequences, we could not predict by what subsequent processes these effects would be produced, nor in what mode they would manifest themselves.

With the qualifications and exceptions,[4] thus stated, I do not see how it is possible, that the pathology of the living economy can be deduced, or inferred, from its physiology; and before proceeding to the chief object of this chapter, I wish to call the attention of the reader to the very limited extent, and the unimportant character, of these qualifications and exceptions. They are more nominal than real. When closely examined and analyzed, they reduce themselves within very insignificant dimensions. Swelled to their utmost possible importance, they hardly amount to anything more than the truisms, that if a part of the body, manifestly intended for the accomplishment of a certain purpose, is injured or defective, then that purpose will in some degree fail of being accomplished; and that, where certain associated and mutually dependent processes are necessary to the production of certain results, a disturbance, or failure, of one of the processes will be followed by a disturbance, or failure, of the others, and by [92] the imperfection, or failure, of the results themselves.[5]

If the doctrine which has been announced is sound and true, and to the extent which is thus asserted, then the entire domain of pathology, vast and various as it is, ought to furnish instances and exemplifications of its soundness and truth. And such, I think, is the case. There is hardly a morbid process, in any organ or tissue of the body, that would not serve my purpose, as an instance and an exemplification of the truth which I wish to exhibit. It will be sufficient, however, to cite a few only of these, for this purpose. Let us look first at that pathological process, or series of processes, which is designated by the term *inflammation*. There is no morbid process, or condition, more common than this; there is none more important; there is none which has been more carefully and thoroughly studied; there is none which is better understood. The appreciable elements of which it is composed; its forms and modifications in different organs and tissues; its causes; its tendencies; its terminations; its results, have been very accurately and closely investigated. Now, I ask, if any attainable knowledge of the healthy action of the parts, in which this process is seated, could, of itself, have led us to a knowledge of that diseased action of the same parts, constituting inflammation? Is there anything, susceptible of being ascertained, in the natural functions of these parts, in the properties, [93] the susceptibilities, the actions of the minute arteries, the minute veins, of the capillary vessels, of the nervous filaments involved in this morbid process, which could have presupposed their liability to this process? Could any knowledge of the former have led, by any course or method of reasoning, independent of observation, to a knowledge, or a prediction, of the latter? Could a knowledge of one have been *deduced* from a knowledge of the other? Most clearly and indisputably not. There is nothing, whatever, in the physiological condition and relations of the parts concerned in inflammation, which could have shadowed forth, or indicated, in the dimmest possible degree, their liability to this condition. By what conceivable process of reasoning — by what imaginable steps of logical deduction — could a knowledge of the former have led us to a knowledge of the latter? Do the natural, the unfelt, the unnoticed actions of these minute vessels and nervous filaments presuppose, in any way, their liability to those numerous and complex processes — the contractions and the distentions of the vessels — the increased, the diminished, the irregular velocity of the blood — the pain, the heat, the secretions — which enter as elements into this morbid condition? Certainly not. What physiological properties of these minute vessels could have informed us of their power, under any circumstances, to separate the fibrine from the other proximate constituents of the blood, or to [94] secrete pus? Certainly none. So far, then, as the phenomena themselves of inflammation are concerned, I do not see how it is possible, that they should be inferred or deduced from the physiological phenomena of the parts with which they are concerned, or in which they are seated; and I think, that an examination of all the other circumstances connected with this morbid condition will serve to elucidate and to strengthen this result. Let us take one of these circumstances, — that of the different degree of liability, in the different organs and tissues of the body, to be affected by this morbid process. This difference is very great. Certain parts and organs are very liable to inflammation; other parts and organs are very little liable to inflammation. Now, is there anything in the physiology of these several parts and organs, — in their natural and healthy offices and functions, — from which, by any *à priori* reasoning, these different degrees of liability could have been ascertained? Why are the lungs so frequently, and why is the spleen so rarely, the seat of this pathological process? Why is acute inflammation of the pia mater, and the pleura, so common, and acute inflammation of the peritoneum, so uncommon an affection? It will not do to say, that these different degrees of liability to this disease can be accounted for by any obvious or appreciable differences in the structure and functions of the organs or tissues, in which it is seated. These differences between the peritoneum and the pleura, for instance, are [95] not sufficiently striking to account for the result. Neither will it do to say, as has often been said, that the degree of this liability is in proportion to the importance and functional activity of the different organs and tissues. I do not know that this importance and activity are any greater in the case of the pleura, than in that of the peritoneum; — I do not know that they are any greater in the case of the lungs, than in that of the kidneys. Let us test the value of this pretended explanation, by a reference to the mucous membrane of the stomach. It would be difficult, I think, to find any part of the body, in which, from mere *à priori* reasoning, we should be justified in looking for acute inflammation more frequently, than in this. In what part is there greater activity of function? In what part are more important processes carried on? In what part is there a quicker or more delicate susceptibility to impressions? What part is more intimately connected with the other important acts and organs of the body? Is it not the great centre of the organic sympathies? What part is more constantly exposed to the action of irritating substances? And yet, notwithstanding all these apparent, and *à priori* causes of acute inflammation, very few tissues, or organs, of the system are so rarely affected by this morbid process as the mucous membrane of the stomach. Certainly, nothing can show more clearly the utter futility of the attempt to explain the fact of which I am speak-[96]ing, by referring it to the differences in the importance and activity of the functions of the different organs, than this striking exemption of the gastric mucous tissue from attacks of acute inflammation.[6]

I will very briefly allude to one other circumstance, connected with inflammation, which will serve still further to illustrate the doctrine, which I am advocating. I mean the different forms, under which this morbid condition shows itself, not only in dissimilar organs and tissues, but in the same organ, or tissue, at different times, and under different circumstances. Sometimes the march of inflammation is rapid; sometimes it is slow. Sometimes, and under certain circumstances, the irresistible tendency of this process is to extend and multiply itself throughout the same, or even throughout widely different, and dissimilar organs and tissues of the body. At other times, and under other circumstances, no such tendency exhibits itself. Now, if it is obvious, as I think it is, that this pathological process, even in its simplest form, and on the supposition, that it never showed itself in any other form than this, could not have been inferred, by any mode of reasoning, from the physiological actions of the parts in which it occurs; still more evident is it, that the various and diverse forms of this process, of which I have spoken, could never have been so inferred; and an examination of all the more obscure and complicated phenomena of pathology will lead to the same conclusion. Why are organic alterations [97][7] of the aortic valves so much more frequent; — in the proportion of nearly twenty to one, — than similar alterations of the valves of the pulmonary artery? What is there in the functions and offices of any portion of the body, from which the existence, the properties, and the tendencies of *tubercle* could have been predicted, or deduced? On what physiological grounds could the predilection of this morbid deposition for the lungs have been anticipated? Why is this deposition almost invariable commenced in the upper portion of these organs? Why is the inferior portion of the lungs more frequently the seat of acute inflammation, than the superior portion? What knowledge of physiological relationships could ever have indicated the existence of those associated morbid actions and conditions, which are found in the exanthematous fevers? What means had physiology by which it could have predicted the connexion between the cutaneous efflorescence, and the inflammation of the fauces, in scarlatina; or that, between another form of cutaneous inflammation, and an inflammation of the mucous membrane of the air passages, in measles; or that, between congestion of the spleen, inflammation of the aggregated follicles of the small intestine, and a peculiar cutaneous eruption, in typhoid fever? What knowledge of the physiological composition, properties and relations of the blood could have informed us, that in all simple, acute inflammations, the relative proportion of fibrine in this fluid would [98] be found augmented; while in many other diseases, in continued fevers, for instance, it would be found diminished? Is there anything in the healthy action of the kidneys, from which we could have inferred their power, under certain circumstances, and by a perverted action, of separating from the animal fluids, sugar and albumin? There can be but one answer to all these questions: and to hundreds of others, of a similar character, which might easily be asked. In no case, with the unimportant and qualified exceptions, which have already been made, can the pathological processes, conditions and relationships of any organ, or tissue, of the body, be inferred or deduced from the known physiological processes, conditions and relationships of the same parts. The knowledge of pathological phenomena does not *flow from* the knowledge of physiological phenomena. The science of pathology is not built upon the science of physiology; the former cannot be deduced from the latter. Each science consists in its own phenomena, and their relations; and these phenomena and relations can be ascertained in only one way, and that is by the direct study and observation of the phenomena and relationships themselves. There is one sense in which a knowledge of the normal structure, and the physiological actions, of the body may be said to be necessary to a knowledge of its abnormal structure, and of its pathological actions. We need the former as a *standard of comparison* for the latter. In order [99] to know what constitutes a morbid alteration of structure, we must know in what the healthy condition of this structure consists; and the same thing is true, of course, of its physiological and of its pathological actions. But this, it seems hardly necessary to say, has nothing to do with the question, which I have been considering in the present chapter.[8]

CHAPTER VII.

Relations of pathology to its causes. Etiology. Our knowledge of the causes of disease the exclusive result of observation. Etiology not to be deduced from pathology. Illustrations. Age. Sex. Season.

THE relations of pathology to all those substances, agents and influences, which act as its causes, which convert physiological actions and conditions into pathological actions and conditions, constitute the science of etiology. I wish to show, that the nature and foundation of this department of the science of life differ, in no degree, from the nature and foundation of those other departments, which have already occupied our attention. With certain unimportant qualifications, our knowledge of the causes of disease is the direct and exclusive result of observation and study of the causes themselves.[1] No attainable knowledge of the phenomena themselves of pa-[100]thology can ever lead us, independent of experience, to a knowledge of the causes of these phenomena. The phenomena themselves can be ascertained only by observation; the same thing is true of all the relations of these phenomena. Let us illustrate the doctrine, thus stated, by a reference to some of these relations; and in the first place, to some that are simple in their character, and well ascertained. There are certain diseases, for instance, which sustain a very definite relationship to certain ages, or periods, of life. That peculiar form of acute inflammation, which has received the popular name of croup, occurs much more frequently during a certain limited period of life, than at any other period. The same thing is true of acute inflammation of the pia mater. A large proportion of both these diseases are found in children, between the ages of two and of seven years. Tubercular depositions in the lungs take place much more frequently between the fifteenth and the thirty-fifth years of life, than at any other period; and the same thing is true of typhoid fever.

Apoplectic extravasation into the brain is much more common after the forty-fifth year of life, than it is before this age. In all these cases, there is nothing in the diseases themselves, which could have led, by any process of reasoning, independent of experience, to a knowledge of their respective relations to certain periods of life. No *à priori* considerations could have led to the conclusion, that any one of these diseases should [101] have been more frequent during one period of life, than at others. Again, some diseases are much more common in one sex, than in the other. In early life, for instance, it is found, that males are more subject to acute inflammation of the pia mater, than females; while females are more subject to hooping cough, than males. Diabetes is much more common amongst males, than amongst females. Now, in all these, and in similar cases, so far as the simple and direct relationship between the disease and sex is concerned, no acquaintance with the diseases, themselves, could have indicated the relationship. The latter could not have been *deduced* from the former. The same remarks may be made in regard to the influence of season in the production of various diseases. Prior to experience, and independent of it, no one could have known, that pneumonia and bronchitis would be most prevalent during one season of the year, and dysentery during another.[2] If all this is true, so far as these simple and well ascertained relationships are concerned, it is quite unnecessary to multiply illustrations drawn from causes of a more complex and obscure character. No pathological process, or condition, can be referred to any agent or influence, as its cause, by any method of reasoning, independent of direct observation of the relationship itself. The latter cannot be *deduced* from the former. A knowledge of one does not, in itself, lead to a knowledge of the other. [102]

CHAPTER VIII.

Relations of pathology to its modifiers. Therapeutics. Rationalists. Empirics. Therapeutics not deducible from pathology. Inflammation. Periodical diseases. Cinchona and arsenic; Relations between them. Action of remedies on disease, not deducible from their action in health. Opium. Cinchona. Calomel. Action of remedies on the human body not deducible from their action on those of other animals.

THE next relationship, the nature and character of which, I have to investigate, is that which exists between morbid processes and conditions, or diseases, on the one hand, and those substances, agents, and influences, on the other, which are endowed with the property of arresting, or controlling, or modifying these processes, or conditions. These substances, agents and influences constitute what has been called the *materia medica*. The science of *therapeutics* consists in their relationships to disease; and their application to their appropriate purposes and uses constitutes the *art* of therapeutics. *Practical medicine* comprehends, and consists of, the phenomena of pathology, and the relations of these phenomena, amongst themselves, to their causes, and to these their modifiers.

Writers upon the science and the art of medicine have always been, so far as the subject now before us is concerned, divided into two classes, or schools; those of the *rationalists*, and of the [103] *empirics*. The former have always been, and still continue to be, the most numerous and powerful. Their doctrines have pervaded and governed the medical world. They claim to be more *philosophical*, than their opponents, the empirics. They profess to be governed and guided, in their theory and practice, by what they are pleased to call *rational principles*. They allege, that their therapeutics is founded upon rational *indications*. They claim, not merely to cure diseases, but to cure them *philosophically*, and in conformity to their *rational principles*. They claim, not merely to have ascertained the relationship, which exists between diseases and their remedies, but to understand the *nature* and the *reasons* of this relationship. They pretend to explain the *mode* and *manner* in which these remedies produce their results. Their doctrine is, that therapeutics is *founded upon* pathology; that the former is deduced from the latter. They are very confident in their knowledge of the intimate *modus operandi* of their remedies. The empirics, on the other hand, deny all this. They say, that the whole science and art of therapeutics are founded upon simple experience. They say, that our knowledge of the relationship between disease and their modifiers, is the sole and exclusive result of observation of this relationship itself. They disclaim any knowledge of the intimate and essential nature of this relationship. They deny, that any acquaintance, however complete and accurate, with [104] the phenomena of pathology, could ever, of itself, have led to a knowledge of the relations, which exists between these phenomena, and those substances and influences in nature, endowed with the property of arresting or controlling these phenomena. They deny, that therapeutics is founded upon pathology. They deny, that by any process of reasoning, the former can be deduced from the latter. This doctrine, I hardly need say, is the doctrine of this essay; and the remaining portion of the present chapter will be devoted to its statement and illustration.[a] [105]

---

[a] It is constantly alleged, by medical writers, that all rational and philosophical practice must be deduced from pathology. Some of these systematic practitioners would seem hardly willing that any disease could be cured, or indeed *ought* to be cured, unless the cure could be effected *rationally*, and according to rule. Mr. Lizars, in a paper of vol. x. of the Edinburgh Medical and Surgical Journal, upon the *nature* and *cure* of acute inflammation, says: "Many, I have no doubt, will contend, that the explanation of the order of these actions and phenomena is of no avail — of no practical utility; that when disease exists, we have a sufficient knowledge of it, and that our aim then should be to cure the malady. But to such reasoning I have only to answer, that on the precise and correct knowledge of the theory of any disease, must depend the treatment. It has been this taking for granted that has impeded the advancement of medicine. Thus, disease is described as it occurred to the practitioner, and his nostrums of treatment detailed; but no accurate theory of the disease is given, and how the remedies did effect, or were likely to effect a cure, is never dreamt of. . . . . I shall now proceed to show how far a correct knowledge of the theory points out the treatment; for I conceive however satisfactorily practice may establish the treatment of any disease, yet, if we do not clearly comprehend its nature, and the operation of the remedies employed, that we still labor in the dark, and are pure empirics." It would be difficult to find a fuller and clearer

If there are any limitations to this doctrine, they are very partial and unimportant. I will allude to one of these limitations, which, however, [106] is more apparent than real. The connexion between diseases and their causes having been ascertained by observation, we might safely conclude, without waiting for the positive knowledge of experience, that if the cause should be removed, the disease would disappear. But this conclusion consists merely in an application to the particular case, or class of cases before us, of the law of the invariableness of relationships. Having ascertained, for instance, that a certain degree of mechanical pressure upon the brain was followed, immediately, by a perversion or suspension of certain functions of this organ, we might safely conclude that, in conformity to the law of the invariableness of relationships, if this pressure should be removed, the perversion, or suspension of the functions of the brain, would no longer exist. But in this, and in all analogous cases, the relationship must be direct and simple.[1] There must exist no intervening phenomenon between one event and the other, to destroy the directness and simplicity of their connexion. Whenever this is the case, the application of the law wholly fails us; and we must remain entirely ignorant of the effects of removing, or destroying, the first link in the chain of relations, upon the last, until we have ascertained, by experience, these effects upon the intermediate links of the chain. [107]

With the qualification thus stated, a qualification, as I have already said, more nominal than real, *all* our knowledge of the relations between diseases and their remedies, or modifiers, is solely and exclusively the result of direct observation. The very existence of any such relationship would be utterly unknown to us, had it

---

statement of the prevalent false philosophy in medicine, than this. Mr. Lizars's theory of inflammation makes the first essential step in the series of morbid processes, consist in disturbance of the *nerves*; and then, by what he calls a process of *rational induction*, he arrives at his treatment, which consists in the application of hot, anodyne fomentations. It is curious to see with what complacency he regards this treatment; so *philosophical* — so readily comprehended and understood!

Again, in the same journal, vol. xxi. a writer, in speaking of the therapeutics of consumption, says: "It is indeed to be regretted, that the unsuccessful results of treatment, suggested by reason and principle, furnish a strong pretext for adopting the bold and blind measures of empiricism; for when rules of science fail, it may be said, can the practitioner be censured for availing himself of those resources, the efficacy of which is demonstrated by experience? This specious argument, we regret to say, has too often been resorted to as a principle of action." p. 160. I cannot well imagine a more extraordinary or monstrous proposition than this. No treatment suggested by "reason and principle," and founded in "rules of science," however disastrous and unsuccessful it results may have been, is ever to be abandoned for any other, "the efficacy of which has been demonstrated merely by experience!" But, monstrous and extraordinary as this proposition is, it is exactly the doctrine of the *rationalists* in therapeutics, divested of its philosophical disguises, and exhibited in its naked and bald deformity. In a subsequent volume, I find the following statement of the same false doctrine: " This relinquishment of theory, however, is impracticable; and every one who knows the constitution of the human mind, is aware, that whatever professions of untheoretical views are given, are necessarily incapable of being realized, and will manifest themselves in one way or another. The human mind naturally clings, in all obscure and unintelligible processes, to something like an explanation; and it is quite as impossible to avoid theorizing about the causes of such processes, as it is impossible to avoid thinking. The man who disavows theory, and especially in medicine, is either a rash, thoughtless, and insane empiric, or is utterly ignorant of what he ought to know well — the laws of human thought — or is at best a hollow and specious deceiver." *Edinburgh Medical and Surgical Journal*, vol. xxiii. p. 181.

not been revealed by experience. No *à priori* reasoning could ever have taught us the *possibility*, even, of arresting, or controlling, the pathological actions of the tissues and organs of the animal system. No conceivable process of logical deduction, unaided by experience, could ever have indicated, or shadowed forth the fact, that the lips of an incised wound could be made to unite, by what is called adhesive inflammation, or that the pain of neuralgia could be relieved by opium.

This doctrine seems to me to be so generally misapprehended, and it is, at the same time, so intimately connected with all practical medicine, that I wish to present it, as fully and as clearly as it is possible so to do, to the reader. With this end in view, let us proceed to examine it, somewhat more in detail. In the chapter on the relations between pathology and physiology, for reasons that were stated, and for the purpose of illustrating the true character of these relations, I referred, particularly, to the well known morbid process and condition, called acute inflammation. For the same reasons, and for the purpose of our present illustration, let us examine the true nature [108] and character of the relationship, between this morbid process and condition, and those substances, agents, and influences, which are endowed with the power of removing, modifying, or controlling it. Let us suppose, that our knowledge of the phenomena of inflammation were such as it is now; and that our knowledge of all its relations, excepting those, which we are examining, were such as it now is, — is there anything in this knowledge, which could lead, in any way, independent of actual experience, to a knowledge of its relations to its remedies, or modifiers? Is there any conceivable process of reasoning, by which the former knowledge could lead to the latter? Could we ever have *deduced* the therapeutical relations of inflammation from its phenomena, or from its relations to its causes? In any rational, or intelligible sense, could the treatment of inflammation have been inferred from its pathology? Does the former flow from the latter? Even if the phenomena of this morbid process, and its relations to its causes, had been much more simple than they are; if it had never presented itself under different forms, in different organs and tissues, and under different circumstances, would this be the case? Looking at the elevated temperature of an inflamed part, we might have been justified, perhaps, in the probable supposition, or conjecture, that by the direct application of cold, we might be able to diminish, [109] or to remove, the morbid heat; and, by this action upon one of the elements of this morbid process, to modify or to destroy the other, and so to mitigate the severity, to modify, or to remove, the disease. But this act of *à priori* reasoning would have consisted, merely, in a conjecture, or supposition, more or less probable. Actual trial of the application itself could alone determine the real relationship between the proposed remedy and the disease. This trial might have shown, not merely that the supposed relationship did not exist, but, on the contrary, that the true relationship was quite different from the supposed one. It might have shown, that the *rational* and *à priori* remedy, instead of diminishing the morbid heat, acted only to increase it; and further experience might have established the fact, that this morbid heat might, under many circumstances, be diminished, or removed, by the application of warmth; all which,

I need hardly say, has actually happened.[b] Positive observation has ascertained, with a considerable degree of certainty, the relations, which do exist between the phenomena of inflammation, and the more or less direct application of cold and of warmth to the seat of the disease. These relations differ very widely, under different circumstances, varying with the seat, the [110] character, the stage, and the complications of the inflammation; and they are such as no method of reasoning, or induction, could ever have ascertained. Again, looking at the accumulation of blood in the tissue of an inflamed part, or organ, we might have been justified, perhaps, in the supposition, or conjecture, that the removal of a portion of this accumulated fluid, from the part, or from its immediate neighborhood, would be followed by a mitigation of the severity of the disease, or by some modification of its phenomena. But in this case, as in the other, the reasoning, if such it can be called, would have consisted merely in a supposition, or conjecture, more or less probable, of the existence of a relationship, which observation alone could determine. This relationship, like the other, observation has, in a good degree, determined; and like the other, also, it is found to differ very widely under different circumstances; varying with the seat, the character, the stage, and the complications of the inflammation; and such as no process of reasoning, independent of direct experience, could ever have ascertained.

An examination of all the other therapeutical relationships of inflammation will render the principle, which I am endeavoring to illustrate, still clearer and more evident. There is not one amongst them, which could have been indicated, even, by any method of deductive reasoning. How could any such reasoning have ever led to the [111] conclusion, that the abstraction of blood from the general circulation would have diminished the intensity, or shortened the duration, or in any way changed the action of this local morbid process? How could any such reasoning have led to a knowledge of the circumstances, in which this abstraction of blood would be followed by beneficial results? How could any such reasoning have led to a knowledge of the relationships, which exists between inflammation, and the operation of calomel, antimony, and opium? Could the effects of these substances have been deduced, or inferred, from any knowledge, however accurate, of the phenomena of inflammation? Manifestly, and indisputably, not. All these effects have been ascertained by simple and direct observation of the effects themselves. It is not possible, in the nature of things, that they could have been ascertained by any other method, or in any other way.

There are certain pathological processes and conditions, one characteristic element of which consists in a distinct and well marked periodicity in their recurrence. These processes and conditions differ very widely from each other in many important particulars; but they agree in this. The most common of these are intermittent fever,[2] and periodical neuralgia. Perhaps there is no therapeutical

[b] Mr. Lizars said, in 1819, "for ten years I have used, invariably, hot, anodyne applications to every acute inflammatory disease, and have never found them fail, in either mitigating or arresting the disease." *Edinburgh Medical and Surgical Journal*, vol. x. p. 408.

relationship better established, than that which exists between these diseases, on the one hand, and cinchona and arsenic, on the other. These substance, when introduced into the sys-[112]tem, are endowed with the power of arresting, or of modifying, the above-mentioned diseases. Is there anything in this periodical element of these diseases, which, by any process of deduction, could have led to a knowledge of its relationship to these substances? Do these substances possess any other known property in common, excepting this of their relationship to these diseases? There can be but one answer to all these questions. No attainable knowledge of the morbid element; no attainable knowledge of these substances, could have ever led, independent of experience, to a knowledge of the relation, which exists between them. Who could have anticipated, that the action of an emetic would relieve the difficult breathing of croup? What *rational* connexion is there between syphilis and the preparations of mercury; or between scrofula[3] and iodine?

It would be a very easy matter to multiply these questions, and to extend these illustrations. Every portion of pathology, and every corresponding portion of therapeutics, would furnish us with material. I hope, however, that I have gone far enough to show, clearly and conclusively, that all our knowledge of the connexion between morbid processes and conditions, on the one hand, and those substances, agents, and influences, which are endowed with the property of arresting, controlling, or in any way modifying these processes and conditions, on the other, is solely and exclusively the result of observation. *Therapeutics is* [113] *not founded upon pathology. The former cannot be deduced from the latter. It rests wholly upon experience. It is, absolutely and exclusively, an empirical art.* There is but one philosophical, or intelligible, *indication*; and that is to remove disease, to mitigate its severity, or to abridge its duration; and this indication never grows out of any *à priori* reasoning, but reposes solely upon the basis of experience.[c] [114]

---

[c] In the early numbers of the Edinburgh Medical and Surgical Journal, there was published a series of anonymous papers, under the title of *The Inquirer*. The subject of No. XVI. of these papers is contained in this question: — "*Does a minute knowledge of anatomy contribute greatly to the discrimination and cure of diseases?*" The paper was suggested by the circumstance, that Dr. Beddoes, in a plan of medical education, addressed to Sir Joseph Banks, proposed that *four* out of *six* years should be devoted principally to *anatomy*! The whole article is compact and solid with the soundest philosophy. I quote from it the following remarks, which although referring particularly to the supposed connexion between anatomy and therapeutics, are still sufficiently applicable to the subject of the text to justify me in transferring them to my pages.

"For our knowledge of the virtues of opium, and cinchona, of mercury, and antimony, we cannot be indebted to the dissecting knife. Observation and experience, grounded generally on accident in the outset, have been the sole foundation of our acquisitions respecting the nature of these our instruments, without which all our anatomical and physiological information were vain. It was not from anatomical considerations, that Sydenham was led to adopt the cool treatment in small pox, that Currie learned the advantages of cold affusion in fever, or that Rollo deduced the utility of animal diet in diabetes. In a word, the greatest anatomists have not been the greatest improvers of medicine, nor among the most eminent of its practitioners. On the contrary, the most distinguished physicians and acknowledged benefactors of the medical art, have not been remarkable for the cultivation of anatomy. Sydenham, Morton, Mead, Fothergill, Home Huxham, Lind, Heberden, Pringle, were not minute anatomists. . . . Let a man be the most correct and minute anatomist, if he have not long and laboriously attended to the appearances and the treatment of diseases, however plausibly he may reason on the processes and

It follows, from what has been said in the foregoing pages, that the therapeutical action of the substances and agents of the materia medica is not to be inferred from their effects upon the body in a state of health. Their pathological relations [115] are not to be deduced from their physiological relations. After having ascertained, that the effect of tartrate of antimony, or ipecacuanha, taken into the stomach, is to excite vomiting, we might, to be sure, independent of experience, have been led to administer one of these articles, for the purpose of removing, from the stomach, by the act of vomiting, any poisonous, or irritating, substance taken into it. So, in cases of disease, attended with long-continued vigilance, having ascertained the power of opium to produce sleep, in a healthy condition of the system, we might be led, by *à priori* reasoning, to the use of the same substance, for the purpose of overcoming the morbid wakefulness. But even in these, and in all analogous, instances, excepting, perhaps, where the action of the article is to remove the cause of disease, as in the case of offending matters in the stomach, just alluded to, or where the action of the article may be strictly chemical, or mechanical, it is only by actual experience, that we can ascertain the effects of the remedies upon the system laboring under disease. It does not necessarily follow, that because opium usually occasions sleep when taken into the healthy system, it will always remove the vigilance of disease. The philosophical reason for this is obvious. Therapeutics consists in the ascertained relations between the substances and agents of the materia medica, and *morbid* actions and conditions of the body; not between these substances and agents, and the [116] *healthy* actions and conditions of the body. And the philosophical reason is sustained by experience. There are many circumstances, in which the morbid wakefulness attending upon disease is not removed, nor mitigated, by opium, in whatever quantity it is administered. Look at delirium tremens. It is now very well settled, that opium has but little effect, in procuring sleep in this disease. And what a rebuke is contained in the action of this remedy, under these circumstances, upon

---

functions of life, and explain their interruptions and modifications, which constitute health and disease, his knowledge will be but the vain speculations of the theorist, he will be practically more ignorant than many an uneducated nurse in an hospital. Let us not mistake the plausibilities of physiological and pathological reasoning, for actual knowledge, for they have their epidemic periods of change; nor let us believe that the curious part of our inquiries are always absolutely useful. . . . . Can a physician be directed to prescribe blood-letting judiciously by knowledge of the particular course of the arteries and veins? or to recommend with skill the administration of purgatives and emetics, by an acquaintance with the structure of the stomach and bowels? Would he not apply, with equal propriety and success, the stimulus of the aspersion of cold water, or the pungency of hartshorn, to a person in syncope, although he were ignorant of the nerves of the skin, or of the Schneiderian membrane? Were pleurisies and peripneumonies more successfully treated, after the arteries and veins of the lungs were described, and their cells injected with quicksilver?" *Edin. Med. and Surg. Jour.* vol.. v. p. 70, *et seq.*

It seems difficult to account for the fact, that such seeds as these should have produced so little fruit, except that they have been choked by the tares of a false *à priori*, and miscalled *rational* philosophy. Why else have not such sentiments taken deeper hold of the British medical mind?

In a letter to Dr. Jenner, dated May 14th, 1806, Thomas Jefferson says, — "Harvey's discovery of the circulation of blood was a beautiful addition to our knowledge of the ancient economy; but on a review of the practice of medicine before and since that epoch, I do no see any great amelioration which has been derived from that discovery." *Baron's Life of Jenner*, vol. ii. p. 95.

our complacent *à priori* philosophy, and our boasted *rationalism* in therapeutics! By what method of what we are pleased to call *rational induction*, could it have been ascertained, that in a disease, strictly functional in its character, not attended with inflammation, and marked especially by nervous excitement and wakefulness, not only would opium be found to be nearly destitute of any power; but, further, that this substance might be given in enormous doses, without producing any perceptible effect, whatever, either upon the disease, or the system generally? Is there anything in the physiological relationships of cinchona, that could have led, without the teachings of direct experience, to a knowledge of its pathological relationships? Does it produce any effect upon the healthy system, which could have indicated, even, in the most indefinite manner, its power of arresting, or controlling, intermittent fever? Calomel, when introduced in moderate quantity, into the system in a state of health, occasions severe [117] local inflammation, attended with general febrile excitement. Is there anything in this action of calomel, which indicates the power of the same substance to arrest and control extensive and intense local inflammation? On the contrary, so far as mere *à priori* reasoning is concerned, would it not have been more philosophical to have concluded, that this new inflammation, with the general disturbance of the economy attending it, would tend to increase, rather than to diminish, the severity of the original disease? The most that can be said in favor of the doctrine, the unsoundness of which I am endeavoring to show, is this; — that, in a few instances, the therapeutical properties of the articles of the materia medica may be, to a certain limited extent, and with many qualifications, inferred from their actions on the healthy functions. But in these instances, the *inference* is only more or less probably; and its correctness can be tested and ascertained only by the results of actual experience. The *inference* is not to be relied upon any farther than as an indication of an experiment or trial; the only foundation of our therapeutical knowledge consists in the result of the experiment or trial itself.[d] [118]

Remarks similar to the above, and for similar reasons, may be made in regard to the effects of the articles of the materia medica upon animals. The action of these substances upon the human body, in a state of health, is not to be positively inferred from their action upon the bodies of other animals in a state of health. So far as the structure and functions of the several organs and tissues of these animals resemble the structure and functions of the corresponding organs and tissues in man, the action of these substances must be the same. But, in many instances, there is more or less difference in the structure and functions of these corresponding organs; and just in proportion to the degree of this difference, will the relations, between the organs and the substances of the materia medica, differ. It is perfectly well known, that some animals, high in the scale of organization, take, with impunity, into their

[d] Sir Humphrey Davy says, in a letter to a young friend, — "I have heard of some experiments you have made on the action of digitalis, and other poisons, on yourself. I hope you will not indulge in trials of this kind. I cannot see any useful result that will arise from them. *It is in states of disease, and not of health, that they are to be used*; and you may injure your constitution without gaining any important result." *John Davy's Life of Sir H. Davy*. Vol. i. p. 104.

systems, substances that are fatal to the life of man. For reasons precisely similar to these, the therapeutical action of substances upon the human body is not to be inferred from the therapeutical action of the same substance upon the bodies of other animals. So far as the morbid actions and conditions of the several organs and tissues, in these animals, resemble the morbid actions and conditions of the corresponding organs and tis-[119]sues in the human body, these therapeutical actions must be the same, according to the great law of the invariableness of relationships, of which I have so often had occasion to speak. But, certainly, in many instances, it may be in all instances, these morbid actions and conditions are not absolutely identical in their character; and just so far as they differ from each other in character, must they necessarily differ in their therapeutical, as well as in their other, relationships. Each class of animals has its own structure; is endowed with its own properties; has been made subject to its own laws; is connected with all surrounding substances, and agents, by its own relationships. This structure, these properties, these laws, these relationships, can be ascertained only by studying them in each separate class of animals, to which they belong, and with which they are connected. It is as unsafe, as it is unphilosophical, to attempt to *infer*, or *deduce*, positively, and independent of experience, those which may exist in one class, from those which are found to exist in another. Analogy may indicate or suggest the direction in which our researches should be carried; it can do nothing more, neither here nor elsewhere; and to this very humble process should its functions always be limited. [120]

CHAPTER IX.

Diagnosis; its importance, and its relations to Therapeutics. Illustrations. Pleurisy; Typhoid Fever.

THE considerations, contained in the preceding chapter, lead directly and obviously to a distinct and clear conception of the nature, the importance, and the relations of diagnosis. Diagnosis is an art, depending upon a *knowledge of pathology*. Just in proportion as this knowledge is positive, accurate, and complete, is our diagnosis positive, accurate, and complete. The two are correlative conditions. The philosophical reason of the practical importance of diagnosis, is simply and manifestly this; — it is the expression of one of the terms in every problem of cure; — it constitutes what may be called one of the elements in every therapeutical operation, or analysis. It is the only term, the value of which it is difficult to ascertain; it is the great element, upon a full knowledge of which, the certainty of every therapeutical operation depends. Therapeutics consists in the relationships which exist between pathological actions and conditions, on the one hand, and the articles and agents of the materia medica, on the other. These relationships, like all others, are fixed and invariable. The properties of the articles and agents of the materia medica are easily ascertained. It is not from any difficulty in ascertain-[121]ing these properties, that the uncertainties of therapeutics arise. These uncertainties grow out of, and rest in, the imperfection of our diagnosis; the incompleteness of our knowledge of pathology.[1] Just in proportion to the perfection and absoluteness of our diagnosis; just in proportion to the completeness of our pathological knowledge, will be the certainty of our therapeutics. All practical medicine depends upon a knowledge of three things, to wit: pathology; the articles or agents of the materia medica; and the relationships between these two elements: and nearly all the difficulties, the obscurities, the uncertainties, the imperfections of practical medicine, grow out of the difficulties, the obscurities, the uncertainties, the imperfections of our pathological knowledge, or, in other words, of our diagnosis. Let us endeavor to illustrate the doctrine thus stated.

For this purpose, it is of very little importance what pathological conditions, or diseases, we make use of. Let us, in the first place, however, choose some one of these conditions, least obscure, and least complex, in its character; and in this respect, there is no one that can answer our purpose better than acute pleurisy. There is no disease of an important organ better known than this. There is none less complicated in its pathology, and in its relations; there is none, the diagnosis of which, in its several stages, and in its different degrees of severity, can be more clearly or positively made out. We can ascertain, with a great degree of accuracy, the [122] seat and the extent of the inflammation. We can follow, with a considerable degree of positiveness, some of the most important changes, and phenomena, which accompany this inflammation. We know very well the condition of the lung, lined by the inflamed membrane, and compressed by the effused fluids; and we can measure, with a good degree of accuracy, the quantity of fibrine and of serum, deposited in the cavity of the pleura, and estimate the variations in this quantity during the different stages of the disease. This inflammation is not often complicated with other serious pathological conditions; and when these complications do exist, they are generally easily ascertained, and their importance easily appreciated. In short, the diagnosis of active pleurisy, in all its elements, is very complete and positive; and in exact correspondence to this completeness and positiveness is the accuracy of our knowledge of its therapeutical relationships. A quart of blood, drawn in a given time, from the arm, will always, under the same circumstances, produce precisely the same effects. Two grains of calomel, or half a grain of opium, or a quarter of a grain of tartrate of antimony and potassa, or the three substances in combination, introduced every three or six hours into the system, will always, under the same circumstances,[2] be followed by precisely the same results. All true and direct relationships are invariable. The circumstances of the system, in acute pleurisy, are susceptible of [123] more accurate estimate and appreciation, than they are in many other diseases; and just in proportion to the accuracy of this estimate and appreciation, is the certainty of our knowledge of the therapeutics, or the treatment, of this disease. If our knowledge of these circumstances could be made perfect and absolute; if it could be made as nearly so as our knowledge of the composition and properties of calomel, tartrate of antimony and potassa, and opium is, then our knowledge of the relations between these circumstances, on the one hand, and these substances, on the other, would become, also, perfect, and absolute. The imperfection of our knowledge of the relationships between these two elements — the disease, on the one hand, and the therapeutical agents, on the other — must grow out of, and depend upon, the imperfection of our knowledge of one of the elements themselves. And here, as everywhere else in practical medicine, this imperfection is in the knowledge of the disease; not in that of the composition and properties of the remedies. This composition and these properties we are sure of; they are positive; they are constant. We may be entirely certain, that the calomel, the tartrate of antimony and potassa, and the opium, which we administer in a case of pleurisy today,[3] are identical in composition and character with those which we administered in another case of the same disease yesterday. And if the two cases of disease were alike, the effects of the remedies [124] must necessarily be the same. The difficulty, and the only difficulty, consists in ascertaining the identity, or the degree of similarity between the two cases of disease. But even in a disease so simple as acute pleurisy, occurring in a person otherwise in a state of entire apparent health, it is difficult, perhaps it would be nearer the truth to say impossible, to find two cases in all respects alike.[4] The obvious and appreciable elements, which are united to constitute the disease, differ in many respects in different cases; and these elements are also constantly changing, in themselves, and in their relations to each other. The state of the system, at the commencement of the disease — *a state, or condition, which is the aggregate result and product of physiological and pathological actions and relations, that have been going on, and have existed, ever since the life of the individual commenced* — must also be widely different in different cases; and in no two, probably, precisely alike. Then, in addition to all this, there are peculiarities in different individuals, less obvious in their character, of a more subtle and recondite nature, and known only by their effects, which would more or less powerfully modify the disease itself, apart from the differences already enumerated.[5] For these reasons, even in that simple form of disease, which I am now speaking of, our diagnosis, in all its elements and relations, can never be absolute and complete; and for this single and simple reason, our therapeutics must par-[125]take of the same character of imperfection. The actual degree of certainty, to which our therapeutics is capable of being carried, and the real extent and power of our remedies over disease, will be made the subject of a separate chapter.

A few remarks, similar to the foregoing, in relation to some morbid condition, or disease, of a more[6] complicated and obscure character, will be sufficient to answer the end of the present chapter. Let us take that disease, which is now generally known in this country by the name of typhoid fever, — the *dothinenteritis* of many French writers, and the abdominal typhus of the Germans. The pathology of typhoid fever is very complicated. Nearly all the functions of the body are more or less seriously disturbed during the course of the disease; and very extensive and numerous structural alterations are found, on examination, in fatal cases. In the present state of our knowledge, the therapeutical relations to this disease are very imperfectly known. It is not yet ascertained, that any of the articles, or agents, of the materia medica are possessed of any considerable power over it. It may be, that there are no articles, or agents, in nature, endowed with this power, to any very positive or great extent. This is a question, which can be settled only by further observation. But be this as it may, and on the supposition, even, of the existence of these substances, endowed with this power, our ability to apply them with success will depend upon the accuracy and [126] positiveness of our knowledge of the disease. Every peculiarity in its pathology, in any given case; every variety in the combination and proportion of its numerous and complex elements; every change in these elements, and in their relative proportions, will necessarily change, in a corresponding degree, the relations between the disease and its therapeutical modifiers.

CHAPTER X.

Diagnosis, twofold: — Nosological and Therapeutical. Elements and means of nosological diagnosis. Diseases not to be required to be wholly unlike each other. *Typhoidal* fever, and *congestion* common elements. Locality of disease. Nature, or character, of disease. Combination and succession of certain phenomena. Symptoms. Relative value of these several elements. Tendencies of modern researches. Therapeutical diagnosis.

DIAGNOSIS is twofold, to wit, nosological, and therapeutical. It is the object of this chapter to point out the character of each of these kinds of diagnosis, and the differences between them. Nosological diagnosis is that to which this term is usually applied; to which, indeed, it is generally confined. Considered as a science, it consists in the *individuality* of each separate morbid process, or series of processes; or of each separate morbid condition; considered as an art, it consists in the power and the act of distinguishing between these several individual processes, or conditions. The [127] number of these separate processes, or conditions, thus distinguished, and individualized, is the number of separate *diseases*, to which the human body is subject: their classification, or arrangement, according to their differences and resemblances, constitutes systematic or methodical nosology: — the names, which are applied to them, constitute medical nomenclature. The elements of all diagnosis are to be found exclusively in pathology and its relations.

The opinions of medical men have always been, to a great extent, confused, indefinite, and contradictory, in regard to the true principles of nosological diagnosis. I am not speaking now of the nosological arrangement of diseases; this subject will be more fully considered in another place.[1] I mean, that there has been no general agreement amongst medical men, in regard to the true principles, and the philosophical foundation, of nosological diagnosis. There has been no common and clear recognition of these principles. Nosologists, and other systematic writers, have differed very widely amongst themselves in regard to what should and what should not constitute a separate disease. Many of them have elevated to this important position a large number of comparatively trifling symptoms, merely, even of a single disease: others have confounded, under the same name, diseases essentially dissimilar. It may be true, that these differences of opinion are, to a certain extent, unavoidable; that they grow out of diffi-[128]culties inseparable from the subject with which they are connected. Many diseases are so complicated in their pathology; they are so frequently constituted, in part, by processes and conditions, which enter largely into other related, but dissimilar, diseases; these diseases approach and touch each other in so many respects, and at so many points, that it may not be possible, always, in the present state of our knowledge, to fix upon positive means, and to lay down positive rules, for distinguishing between them. Let us endeavor, however, to do this, as far as the actual state of science, and a correct view of the subject, will enable us. Let us endeavor to ascertain the true character; to enumerate and to appreciate the legitimate means and elements of nosological diagnosis; to see, as far as is possible, in what the identity and dissimilarity of individual diseases consist.

Before proceeding to do this, and as a preliminary step to our endeavor, let us notice one condition, or circumstance, of a negative character, — or which, in other words, *ought not* to be recognized and admitted as an element of diagnosis. I mean, that diseases must not be required to be *utterly* and in *all respects*, unlike each other, in order to constitute them distinct, individual species. There are several very important morbid conditions, which are common to a large number of separate diseases; a circumstance which necessarily deprives these conditions of any considerable degree of diagnostic value. This is especially [129] the case with that series of morbid actions, which we call *fever*. This term is strictly *generic*, and it ought always to be so used. This associated phenomena, to which this term is applied, cannot, with any propriety, be said to constitute a specific, or individual, disease; they enter as elements merely into the composition of a great number of separate and widely dissimilar diseases. Inflammation of the pia mater; inflammation of the kidneys; inflammation of the pericardium, — are all alike attended by fever; and there is nothing, whatever, in the character of this latter element which distinguishes one of these disease from the rest. The fever, then, is strictly common to them; and so far as their diagnosis, amongst themselves, is concerned, is of no value. Furthermore, this morbid condition, which we call fever, may be marked by certain very prominent and striking peculiarities, and still remain of very trifling importance as an element, or means, of specific diagnosis. There is one form of fever, which is called *inflammatory*: it usually accompanies acute inflammations of an open, frank, or sthenic character; and is marked by a strong, hard pulse, moderately hot skin, thirst, a moist, whitish tongue, and no very striking degree of muscular debility. This form, as has already been said, is present in many separate diseases. There is another form of fever, to which the terms *typhoid*, or *adynamic*, or *asthenic*, have been applied, and which differs in many respects from the former. [130] Now what I wish to say is this, — that the presence in fever, even of these strongly marked peculiarities, still fails to impart to the fever any considerable value in nosological diagnosis. The inflammatory, or sthenic, form of this morbid condition attends many dissimilar diseases; and the same thing is true of the typhoid, or asthenic, form. The latter is usually present in the diseases, which have received the names of typhoid and typhus fevers; and it also very frequently accompanies small pox, scarlatina, some forms of pneumonia, and other local affections. This typhoid element, thus common to many diseases, unlike each other in several or in all other circumstances, cannot, certainly, be regarded as an element of great or primary importance in diagnosis. The diagnosis of these several diseases, thus marked by the presence of this common condition, must rest upon other circumstances peculiar to each.

Remarks, in every respect similar to the foregoing, may be made in regard to that obscure, but most grave, morbid condition, to which the term *congestion* has been applied. This condition, marked especially, by great disturbances, or rather by an entire loss, of what may very properly be called *the balance of the circulation*, and by profound but unknown modifications of innervation, is frequently witnessed at the onset, or in the early stages, of diseases; while the typhoidal form of fever more commonly shows itself, during their progress, or in their later periods. The former [131] morbid condition, like the latter, may be present in many dissimilar diseases, and cannot, for this reason, be relied upon as a means, or instrument, of diagnosis. It accompanies Asiatic cholera, malignant intermittent and remittent fevers, the grave forms of scarlatina, yellow fever, plague, and so on; so that the diagnosis of these diseases, and of others, under similar circumstances, must depend, not upon this common element, but upon others, with which it is associated. The foregoing considerations are sufficient, — although many others of a similar character might be added to them, if it were necessary, — to show, that different diseases may possess certain very prominent and important elements *in common*, without hindering, in any degree, their separation into perfectly distinct, individual species. They may also agree, in many respects; in regard to their causes; their march, and duration; their relations to remedial measures; and in other respects, and still be susceptible of clear and positive diagnostic distinction. It is quite evident, then, that diseases must not be required to be *wholly* and *in all things* unlike each other, in order to constitute them distinct species.

The positive elements of nosological or specific diagnosis are quite numerous; and they vary very widely in their number and character, in different diseases. They are to be found, as in the case of plants and animals, in *all* the phenomena and relationships, which unite to make up the natural [132] history of diseases. In some cases, they are numerous, complicated, and less positive in their character. They may be found in the seat, or locality, of disease; in the nature, or peculiarity, of the essential lesions in which it consists; in certain symptoms or combinations of symptoms; in its relations to its causes and its modifiers; or in several, or all, of these circumstances, variously united.

One of the most common divisions of diseases; one that has been almost universally recognized, is that which separates them into two classes; — those which are *local*, and those which are *general*. This division, let me observe, cannot be regarded as absolute. Some diseases are much more circumscribed in their extent, and much more limited in their actions and influences, than others; there is a very wide and manifest difference in this respect; but, still, it is not easy to show, that any disease is absolutely local, on the one hand, or absolutely general, on the other. In the simplest cases of local disease, there may be more or less complexity of pathological action; in those diseases, which are regarded as most general in their character, there are tissues and functions of the body, which, so far as we have means of ascertaining, are in a healthy condition. But, notwithstanding all this, the anatomical locality, or situation, of many diseases constitutes one of the chief, and fundamental elements in their nosological diagnosis. There are many diseases, the pri-[133]mary and essential seat of which is in certain organs, or tissues of the body; in which their processes are carried on, and to which they are mostly confined. This circumstance is, of itself, and independent of other circumstances, sufficient to settle, so far as it goes, the nosological diagnosis of certain diseases. If these diseases agree in all other circumstances, but differ in this, of their anatomical locality, they are different, and dissimilar, diseases. This circumstance alone establishes their nosological diagnosis, and fixes one element, at least, of their nomenclature. Inflammation of the kidney is not the same as inflammation of the liver; hemorrhage from the vessels of the brain constitutes one disease, hemorrhage from those of the lungs constitutes another; dropsy of the pericardium is not the same disease are ascites. This locality may have reference to an entire organ, or to the anatomical elements, or tissues, which enter into the composition of the organ. Thus, inflammation of the internal lining membrane of the heart constitutes one disease, and inflammation of the external lining membrane constitutes another. This is one of the simplest and most positive elements of diagnosis, in all cases where the locality, and the primary character, of the disease are sufficiently manifest and certain. There are other cases, in which the localization of a morbid process in any given organ, or tissue, may constitute only a secondary means of diagnosis, or in which it may be rejected almost entirely. This will hap-[134]pen where the peculiar *nature* and *tendencies* of the morbid process constitute its fundamental and most important element; and where its seat, or locality, is of secondary or accidental value. Thus we may have *tubercle*, or *cancer*, constituting, each an individual, and identical disease, in whatever organ, or organs, of the body, it may be mostly, or exclusively, situated.[2]

In the second place, we find an important element of nosological diagnosis in certain characteristics of disease, independent of its anatomical locality. The same organ, or tissue, may become the seat of morbid processes, and conditions, differing, so far as their phenomena and relations enable us to judge, essentially from each other. These differences may be radical and absolute, in the *nature* of the morbid process itself; or they may depend upon the combination and the relations of different morbid actions in the different elementary tissues of the organ, or part, which is the seat of disease. Thus, there may be many distinct and separate diseases in the same organ or tissue. The kidneys, for instance, like most other parts of the body, are subject to acute inflammation, constituting a well-marked, distinct, individual disease, characterized by its own appropriate phenomena; and called, in classical nomenclature, *nephritis*. Again, the same organs are subject to another morbid process, the results of which show themselves, locally, in a great augmentation in the quantity, and in certain striking [135] alterations in the quality, of the peculiar secretion of these organs; which changes are also attended with other pathological conditions of a more general character, constituting all together another distinct, well-marked, individual disease, to which we give the name *diabetes*. Once more, the same organs are subject to still another morbid process, characterized by its own peculiar phenomena, both local and general; differing from either of those already mentioned; and constituting a third distinct, well-marked, individual disease, which is called *albuminuria*, or *Bright's disease*. Softening of the cerebral substance constitutes one disease; an extravasation of blood into it, another.

In the third place, a very important element of systematic diagnosis is to be found, not so much in the principal locality, or the peculiar nature, or character, of the disease, as in a certain *combination*, and *succession* of morbid processes and conditions. Many of the diseases, belonging to this class, possess certain features, more or less prominent or striking, *in common*, which give them a *family* resemblance; but each individual member of the group, or family, to which it belongs, is marked by certain traits, or by some peculiar combination of features, which distinguishes it from the others to which it is allied. The character and value of the diagnostic element of which I am now speaking, may be very clearly shown by a reference to the exanthematous fevers. The [136] most common, and the most important, of these are, small pox, scarlet fever, and measles. What are our means of distinguishing, nosologically, between these several diseases? In what are the elements of our diagnosis to be found? Not in any anatomical locality of either of the diseases; not in any ascertained peculiarity in the nature, or character, of the morbid processes, or conditions, in which they consist. In each of these diseases, our diagnosis depends upon, and consists in, a certain combination and succession, or series, of morbid processes and conditions, characteristic of the individual disease in which they occur. In small pox, these diagnostic elements are to be found, principally, in a series of morbid processes, which take place in the skin. This series of processes is not found an any other disease. It consists in an eruption of a well defined, and peculiar character, commencing at a definite period after the occurrence of other morbid phenomena, and going through a regular succession of changes. This eruption, thus constituting the fundamental element of specific diagnosis, is associated, as has just been intimated, with certain other phenomena, more or less characteristic of this particular disease. Amongst these, are the specific nature of its cause, the determinate duration of the several stages, or periods, of the disease, and its peculiar relations to another allied affection, — cow pox. The diagnosis of small pox consists in the presence of *all* these *associated* [137] phenomena, and in their more or less regular *succession* and *development*. Similar principles of diagnosis are applicable to all the exanthemata, — to the several forms of periodical, and continued, fever, to plague, and to some other diseases. In all these cases, we rely, for our diagnosis, upon the combination and succession of certain phenomena, more or less numerous and characteristic, and differing from each other in the several individual diseases.[3]

There is another class of diseases, the positive, diagnostic elements of which consist almost entirely, or nearly so, in certain symptoms, — the nature of the diseases, and in many cases their causes, also, being wholly unknown. Amongst these, may be mentioned, as types of the class, epilepsy, tetanus, chorea, hydrophobia, and delirium tremens. Each of these diseases is distinguished from the rest of the same family, and for still stronger reasons, from all other diseases, by certain peculiar and characteristic symptoms, and by these alone, or in connexion, as in the case of the two last-named affections, with their specific causes.

The elements of diagnosis, which have been thus indicated, must be definite, fixed, and constant, — each single element, or combination of elements, constituting the diagnostic marks of a given, individual disease, not being interchangeable with those of any other individual disease. Separate and distinct diseases may exist together in the system; and in consequence of this coexist-[138]ence, they may be somewhat modified in their character, and manifestations; but they cannot properly be considered as *convertible into each other*. They may approach each other very closely, or become quite identical, so far as their analogous or common elements are concerned; but their diagnostic conditions must not be subject to this mutual conversion, or blending together. The exact value of these several conditions, or elements, — actual and relative, — is a matter, not susceptible, perhaps, of very positive determination; but I will venture to remark, that amongst those, which are most absolute and distinctive, is the seat of local diseases, and the presence, in those of a more general character, of some obvious, and peculiar anatomical lesion, like that of tubercle, and cancer, the pustular cutaneous eruption in small pox, and the follicular ulceration of the intestines in typhoid fever.

Such I believe to be the fundamental and true principles of nosological diagnosis; by the application of which, the individuality of all diseases, and their character, as distinct species, are to be determined. It is important to observe here, that our ability to apply these principles successfully has nothing, whatever, to do with the soundness of the principles themselves. This ability will depend upon the knowledge, the sagacity, and the skill of the individual observer. The existence of individual diseases is one thing; the power of ascertaining this existence is another: — [139] the former is not dependent upon the latter.[4] It is only within a few years, that we have been furnished with means of distinguishing, with clearness and certainty, between pleurisy and pneumonia; but these two diseases have always been as distinct from each other as they now are. Besides this, it should never be forgotten, that almost all disease are occasionally so impressed and modified, by inappreciable or unknown influences, that their usual diagnostic signs are wanting, or very much obscured, — the diseases being *latent*, as it is called. Cancerous disorganization of the stomach, in some instances, gives no indication of its existence, sufficiently distinct to render its detection possible, during life, even by the most competent and careful observers: and the same thing is true in the case of most other diseases.[a] [140]

---

[a] In my *History of Typhoid and Typhus Fevers*, after stating that there are few general diseases, susceptible of a more certain and positive diagnosis, than the former, I added the admission, that cases might sometimes occur, so enveloped in obscurity, as to baffle the skill of the most careful and

Let me add, in conclusion, that the tendency and result of that accurate, minute, and comprehensive study of disease, which distinguishes the school of modern medical observation, and which marks the advent of a new era in our science, have been altogether and uniformly in favor of a nicer and more positive discrimination between diseases, than has heretofore existed. The opposite tendency, especially in Great Britain and in this country, has been principally owing to the vicious spirit — so generally prevalent, and so potent in its influences — of gratuitous and unwarrantable generalization; — a spirit which gave birth to the preposterous dogma of the absolute unity of all disease, and which led Dr. Armstrong,[5] Dr. Boott,[6] and many others, equally distinguished for learning and ability, to advocate the doctrine [141] of the essential, specific identity of all the various kinds of continued and periodical fever — of typhus, of typhoid, of intermittent, bilious remittent, congestive, and yellow fever. It can hardly be owing to anything else, than the influence of this disposition, that the great majority of British physicians of the present day refuse to admit, *or to endeavor to ascertain*, even, by a thorough and impartial investigation of the subject, the true distinction between the two great forms of continued fever — a distinction that was clearly recognized by such men, amongst their illustrious predecessors, as Huxham,[7] Darwin,[8] and Pringle.[9, b] [142]

---

experienced observers, — that the disease might occasionally be so nearly latent, or so poorly defined, as to be overlooked or mistaken. In a somewhat ungracious review of my book, in the Edinburgh Medical and Surgical Journal, this admission of the difficulty, or impossibility, in a few exceptional cases, of distinguishing typhoid fever from other diseases, and especially from its allied affection, — typhus fever, — is gravely cited as sufficient proof, that the two diseases, — typhoid and typhus fever, — cannot be distinct species! Let me add here, that this question, of the essential likeness, or unlikeness, of these two diseases, — one of the most important and interesting questions of specific diagnosis, that has ever occupied the attention of physicians, — if submitted to the test of the principles which I have laid down, and fairly tried by them, — cannot fail, I think, to be settled in favor of the doctrine of their fundamental dissimilarity. The two diseases will be found to approach each other, very closely, in the possession of those morbid processes and phenomena, — I mean general fever of the typhoidal type, certain changes in the composition and quality of the blood, and certain nervous symptoms, — which are common to many diseases, and, for this reason, of but small value as diagnostic or distinctive characters; while they are separated clearly and broadly from each other, by the presence in one, and the absence from the other, of very strongly marked and constant anatomical lesions, and of groups of symptoms, equally striking, constant and characteristic. Any principles of diagnosis, or any rules of reasoning, that make true typhus fever, and typhoid fever essentially one specific disease, will make small pox and oriental plague, also, nothing but varieties, or modifications, of the same single disease. This result will be found to be absolutely unavoidable.

[b] The opinion of Huxham upon this subject has often been quoted, and is well known. Those of Darwin, Dr. Vaughan, and Sir John Pringle, are probably less familiar to most of those who may be my readers; and I cannot forbear citing the authority of these English observers of the last century, in support of the view stated in the preceding note. In a letter from Dr. Darwin to Dr. Lettsom, dated Derby, October 8th, 1787, there is the following passage: — "If your society proposes questions, I should wish to offer for one, 'Whether the nervous fever of Huxham, — or fever with debility, without petechiæ or sore throat, of flushed countenance, or pungent heat, — be the same as petechial fever, or jail fever?' The former of these, viz., the nervous fever of Huxham, prevails much over all the country at this time." *Life and Correspondence of Dr. Lettsom*, vol. iii, p. 118.

Dr. Vaughan, of Leicester, in a letter to Dr. Lettsom, dated July 27th, 1783, in reference to the same subject, says: — "There is surely a peculiarity in the species of fever you had the goodness to send me and account of, protracting itself to such a length as thirty-five or forty days; it certainly agrees very much

I mean, by what I have chosen to call therapeutical diagnosis, the distinction between individual diseases, or morbid conditions, depending [143] upon the relations of these to the articles and agencies of the materia medica. Nosological

---

with Huxham's Febris Nervosa, which, *notwithstanding Dr. Cullen, is a very different disease to the Febris Carcerum, in its attack, progress, termination, and cure.*" *Ibid.* vol. iii. p. 161.

The testimony of Sir John Pringle to this point is much fuller. "In the description," (*Observations on the Diseases of the Army, Phil. Ed.* P. 298,) he says, "I have endeavored to distinguish them" — malignant or pestilential fevers — "from all others, as far as I could do it, in distempers whose symptoms are so much alike. The nervous fevers are frequently accompanied with miliary eruptions, which have no resemblance to the *petechiæ*; nor have I ever happened to see miliary eruptions in the malignant kind." In reply to some strictures of De Haen, (*Ibid.* p. 384,) he says, still more explicitly: — "I have never considered the jail or hospital fever, and the miliary fever" — meaning the low, nervous — "as similar; and, indeed, I may venture to say, that, as the symptoms of the two are so much unlike, they ought to be treated as different *in specie*; and, consequently, that neither the theory nor the practice in the one ought to be regulated by analogy from the other." Again, he says: — "I have therefore all along considered the jail, or hospital fever — in regard to others that commonly occur in these parts — as a fever *sui generis*, at least as different from either the scarlet, the miliary, or any other eruptive fevers, which are known." *Ibid.* 385.

The strictures alluded to above, by De Haen, had reference, particularly, to the treatment of fever by Huxham and Pringle. De Haen charged these glorious old British observers — the types and ornaments of a school never since surpassed by their countrymen — with bad practice; with a too stimulating and incendiary method in the management of fever. Pringle, in his reply to De Haen, says expressly, that the fever treated by the latter at Vienna was of a different kind from that treated by himself; and in a note to this reply, he makes the following very interesting remarks, in regard to the dissimilarity of the cutaneous eruptions in the two diseases. "After publishing what is above, relating to the distinction, which I conceived was to be made between De Haen's *petechiæ* and mine, I was confirmed in my opinion by Dr. Huck, who, in the year 1763, was at Vienna, and was favored with admittance into all the hospitals there, and in particular had the satisfaction of attending Dr. De Haen himself, and seeing, with that celebrated physician, some of his patients in that very fever, which he calls *petechial*. Dr. Huck examined those spots in Dr. De Haen's presence, and assured me, that they had hardly any resemblance to those which I have called *petechial*, and which he himself had so often seen in the hospitals of the army; but that they were so like flea-bites, that he was apt to believe, that one must be often mistaken for the other." (*Observations on the Diseases of the Army*, p. 384.) Let me say here, that I do not know anything in the annals of medical polemics, imbued with a finer temper, or a more philosophical spirit, than this reply of Pringle to De Haen. It is every way equal — and there can be no higher praise than this — to Louis's defences against the attacks of Broussais and Bouillaud. In place, or out of place, I cannot forego the pleasure of gracing a page of my book with the following passages — truly, words of wisdom, "fitly spoken — like apples of gold in pictures of silver" — from the reply of Pringle: "In fine, Dr. De Haen may be assured, that the regimen, which I propose, stood at first on no other foundation than experience, after my having seen the bad effects of a contrary method, whether by too large or too frequent bleedings in the beginning; or by giving hot things too early, in order to raise the pulse, when it began to sink, or to force a crisis before the common period of the disease. Some of the medicines are superfluous, but I am pretty sure, that none of them are hurtful. . . . . But having once got into a method, which brought about as many cures as seemed otherwise consistent with the circumstances of my patients, lying in a foul air, amidst a constant noise, and often neglected by the nurses, I did not attempt to reduce my practice to more simplicity, than what is mentioned. Yet whatever confidence I may have in the directions, which I have published, I am still ready to alter any part of them, upon a fair representation from those, who have had equal opportunities with myself of seeing and treating this fever. But to oppose either mere theory, *or analogy from other fevers, where the similarity is so disputable*; or to oppose some general maxims from Hippocrates or Sydenham to the observations, which I have offered, as the result of a long and painful experience in a distemper, that no physician could well know but in such circumstance as mine, is a manner of writing, I must say, more fitted for disputations in a school of medicine, than for the instruction of a practical physician." *Observations on the Diseases of the Army*, p. 395.

diagnosis constitutes one of the elements of therapeu-[144]tical diagnosis, but the latter includes, also, many other elements in addition to this. The first condition of therapeutical diagnosis is a knowledge of the individual disease; but many other, and frequently much more important, conditions of this diagnosis, are to be found in other circumstances. Amongst these may be mentioned, for the purpose of illustrating my meaning, the following, to wit; — the extent, and severity, of the individual disease — its period — in many cases, its occurrence in a sporadic or an epidemic form — the age of the patient — and the general condition of the patient previous to the attack of the individual disease.[10] These circumstances do not enter into our nosological diagnosis; but they frequently constitute altogether the most important elements in the therapeutical relationships of disease. The nosological diagnosis of acute pneumonitis, confined to the lower portion of a single lung, does not differ from that of the same disease, involving the whole of one lung, and half of the other; but this difference in the extent of the disease will affect very essentially it therapeutical relationships, and the diagnosis depending upon these. And the same thing is true of the other circumstances, which have just been enumerated — the period of the disease, its sporadic or epidemic form — the age of the patient — his condi-[145]tion at the time of the attack, and so on. Each of these circumstances had an important bearing upon the therapeutical relationships of disease; and the latter will be influenced by every modification of the former.

The paramount importance, in practical medicine, of as complete and positive knowledge, as is attainable, of all the circumstances, which can influence diseases, so far as the effects of remedies upon them is concerned, is so obvious, that I need not insist upon it. This knowledge is, indeed, in many cases more absolutely essential to the safe and proper management of disease, than nosological diagnosis itself. [146]

CHAPTER XI.

PROPOSITION THIRD

AN ABSOLUTE LAW, OR PRINCIPLE, OF MEDICAL SCIENCE CONSISTS IN AN ABSOLUTE AND RIGOROUS GENERALIZATION OF SOME OF THE FACTS, PHENOMENA, EVENTS, OR RELATIONSHIPS, BY THE SUM OF WHICH THE SCIENCE IS CONSTITUTED. THE ACTUAL, ASCERTAINABLE LAWS, OR PRINCIPLES, OF MEDICAL SCIENCE ARE, FOR THE MOST PART, NOT ABSOLUTE BUT APPROXIMATIVE.

The character and conditions of principles in medical science. These principles approximative, and not absolute. This approximative character fixed and determinate. Its degree of fluctuation confined within certain limits. Illustrations.

Proportion of sexes at birth. Law of great numbers. Calculation of probabilities. Laws or principles of therapeutics; their complexity; difficulty of ascertaining them. Gavarret. Conditions of these laws. Facts must be comparable. True value of therapeutical experience. Mistaken notions.

THE constituent elements of a law, or principle, in the science of life, do not differ from those of a law, principle, in physical science. I mean by this, that in the former case, as truly as in the latter, the law consists in the constancy of a phenomenon, or the invariableness of a relationship; or in the nearest possible approximation to this constancy and invariableness, and in nothing else. The law, or principle, is not an element lying back of the phenomena and their relation-[147]ships, or interposed between them, or superadded to them, by any act of reason: — *it consists in the phenomena and their relationships, and is identical with them*; — it is the expression, merely, of these phenomena and relationships, generalized and classified.

But, notwithstanding this essential agreement in the nature and composition of these two classes of laws, there is one fundamental difference between them, which it is necessary fully and clearly to exhibit. With certain limited exceptions, the laws of physical science are *positive* and *absolute*, both in their aggregate, and in their elements, — in their sum, and in their details; but the ascertainable laws of the science of life are *approximative* only, and not absolute. This difference I have called fundamental; it runs through almost the entire science of life, and impresses upon its phenomena, and its laws, peculiarities, which require to be fully developed, and thoroughly understood. To aid the reader in the accomplishment of this desirable object, — to point out and illustrate the true character of these laws, — the conditions of their legitimacy and their value, and the true methods of arriving at them, — is the object of the present chapter.

I have already said, that in physical science, all genuine and direct relationships are invariable. This is as true in the science of life as in physical science; but there is this great difference in the two cases. In the latter, these relationships are, [148] for the most part, susceptible of such analysis, and separation from each other, as to be ascertainable in their singleness and simplicity; in the former, they are, almost universally, so numerous, and complicated, so involved and so intricate, as to defy all such analysis and isolation; and it is this circumstance that gives to the laws of the science of life the peculiar character of which I am speaking.[1] The sum of the phenomena and relationships, in any and in every given instance, is not *positive* and *constant*, but *contingent* and *variable*. This character and peculiarity of the *elements* of the law are, of course, extended to the *law itself*; rendering it, as I have said, approximative only, and not absolute.

But this contingency, or variableness, is not indefinite and unbounded; it is confined within certain limits; and these limits are susceptible of very accurate measurement. Within these limits, the law becomes absolute; their extent determines the degree of its possible fluctuation, or variableness. It is to the existence of this appreciable and ascertainable limitation, that we are indebted for the *comparable* character of the facts, and relationships, which constitute the elements, or materials, of our laws. These facts and relationships, are not *identical*, one with another, but their resemblances are sufficiently fixed to render them available as positive data, in the prosecution of our researches. For instance, one important series of these facts, and relationships, is consti-[149]tuted by *individual life*, — or the sum of the organization and its functions, with their relations, in the individual. Now, although this sum, or aggregate, of phenomena and relationships, constituting one individual, is never absolutely equivalent to the similar sum, or aggregate, constituting any other individual, still the difference between them never surpasses certain determinate limits; the resemblances between them are sufficiently constant and fixed to render them *comparable elements*, and to give them a character sufficiently definite, to constitute them legitimate data for scientific comparison and study. Thus, the continuance of the functions constituting life, in the several classes of vegetable and of animal being, although contingent and variable, in each individual; and not susceptible, in any given instance, of being certainly known, in advance, never exceeds certain limits; and its *average period*, for each species, is ascertainable with great accuracy. The distribution of births between the two sexes constitutes a law of physiology of very great positiveness and uniformity, the individual elements of which are altogether contingent and uncertain; and the positiveness of the law depends upon the fact, that this contingency is strictly confined within certain limits. The same thing may be said of the number of births to each permanent union of the sexes by marriage. This number, in any given instance, is entirely uncertain; but the uncertainty is always limited in degree, so that each single fact, consti-[150]tuted by this variable number, is still sufficiently fixed and definite, to render it subject to comparison with other similar facts, and so to convert it into a legitimate element of a law of the science of life.

Similar remarks may be made in regard to the phenomena and relationships of pathology. Every law, or principle, of pathology consists solely in a generalization of certain phenomena, or relationships. These phenomena and relationships, in each individual of a class or series, constitute a sum or aggregate of uncertain and variable quantity; and the law, which results from their generalization, must partake, in some degree, of this character. But the degree of this variableness, both in the individual sum, or aggregate, and in the whole of these, classified and arranged, constituting the law, or principle, is confined within certain limits, susceptible of being ascertained and measured. This limitation gives to the individual facts a character sufficiently fixed and determinate, to render them susceptible of being compared with each other, and so to convert them into legitimate elements, or constituents, of a law. Were it not for this circumstance, there could be no such thing as science in pathology. There is, for instance, a certain number of phenomena and relationships, the sum of which constitutes a disease, to which we give the name of *pleurisy*. This sum or aggregate is not absolute, and uniform, but contingent, and variable. No [151] one of these aggregates, constituting the disease, is ever exactly equivalent to another; no two cases of pleurisy are ever precisely identical. Still, the differences between them are not unlimited and indefinite; they are always confined within certain degrees. The resemblances between these individual aggregates are sufficiently fixed and positive, to render them determinate and comparable facts;

capable of being used as data, and dealt with in our researches and generalizations, subject to the qualifications already made, as we deal with the data of physical science. The same thing is true of all the other groups of morbid phenomena and relationships, constituting the various individual diseases of the nosology. The sum of these phenomena is more variable and fluctuating in some groups, than in others; our knowledge of these phenomena is more accurate and extensive in some groups, than in others; but the degree of fluctuation is always confined within certain limits, which are susceptible of determinate measurement.[2]

Let us now endeavor to see *by what method*, these individual facts, phenomena, and relationships, can be generalized, so as to constitute the laws or principles of the science of life. Let us see how rigorous and positive this generalization of contingent and variable, but still comparable, facts, can be made, — by what process it is to be accomplished, — and what the conditions are, to which it is subject. I have spoken of the law of [152] the distribution of births between the two sexes. What is this law? and how is it ascertained? Certainly, nothing can be more doubtful or contingent, in any single instance, than the birth of a male or a female child. One event is almost as likely to happen as the other. And even where the number of births is considerably increased, the relative proportion of the sexes is a matter of very great uncertainty. Large families of children are sometimes born of the same parents, consisting exclusively of either one sex or the other; and very frequently the proportion between them is utterly unequal; so that the whole matter might seem to be one of unlimited chance and uncertainty. During the first three months of 1843, the whole number of children born in the obstetrical department of the Philadelphia Dispensary was forty-five: of these, twenty-nine were males, and only sixteen, females; the difference in favor of males being almost equal to the proportion of two to one. But as we extend our investigation, we shall find this difference gradually diminishing, until, at length, the true law of this proportion of the sexes at birth is seen gradually evolving itself from the study and analysis of a *great number of facts*. The number of legitimate births in Paris, during the year 1836, was 19,309. Of these, 9,785 were male; and 9,524 were female: the male births being in proportion to the whole, as 5068 to 10,000. The whole number of legitimate births in France, during the year 1825, was [153] 904,594. Of these, 468,151 were male; and 436,443 were female; the male births being in proportion to the whole, as 5175 to 10,000.[a] *But this average result is not to be taken as the positive and absolute expression of the law before us*. The result is still subject to a certain degree of variableness, or fluctuation; the amount of which can be ascertained by an arithmetical process, the elements of which are to be found in the numbers themselves, and which is known as the *calculation of probabilities*. The result of the application of this process to the two illustrations, just given, is as follows. In the first instance, although the positive result showed the chance of a male birth to be 0.5068, a calculation of the probabilities shows, farther, that this chance may vary, in either direction, above or below the observed result, to the

[a] Principes Généraux de Statistique Médicale. Par Jules Gavarret. p. 76, et. seq.

extent of 0.0102: — so that the law derived from these numbers would be, not that the chance of a male birth in Paris, during the year 1836, was rigorously as 5068 to 10,000; but that this chance varied between 5,170, and 4,966 to 10,000: or, that it might have been considerably more, or slightly less, than even, or equal. In the second instance, although the positive result showed the chance of a male birth to have been 0.5175, an application of the calculation of probabilities shows, further, that this chance really varied [154] in both directions, above and below the observed result, to the extent of 0.0015; so that the law derived from these numbers would be, not that the chance of a male birth in France, during the year 1825, was rigorously as 5175 to 10,000; but that this chance varied between 5,190, and 5,160, to 10,000. It will be noticed, that the extent of the fluctuation is very much less in the second, than in the first instance; and the reason of this is to be found in the vastly *greater number* of facts, constituting the law. The law of proportion between the sexes, at birth, in France, during the year 1825, is absolute, within the limits, thus ascertained, by an application to the observed data of the calculation of probabilities; and the law approaches absoluteness and invariableness, just in proportion to the multiplication of the data, or facts, from which it is derived, and by the analysis and generalization of which, it is constituted; and although it may never, from the very nature of its elements, acquire the positive character which belongs to many of the laws of physical science, the degree of its uncertainty may be rendered almost indefinitely small, and so unimportant, as to be practically disregarded.

There is one condition of the legitimacy of the laws, or principles, resulting from the process, and established by the methods, just described, obvious enough, to be sure, but which it may be well to point out and to illustrate. This condition is — in the words of M. Gavarret[3] — that the *sum,* [155] *or aggregate, of possible causes* of the facts, which constitute the elements, or materials, of the law, must remain the same. When this condition fails, the law will be modified, in correspondence with the new element, which has been introduced into the sum of the possible causes of the facts, or phenomena, with which it is concerned. And such a modification, when it exceeds in extent the limits of variation within which the law may oscillate, is to be taken as evidence, that the sum of possible causes has changed, and that some perturbating element has been introduced amongst them. Thus, during the years 1824 and 1825, the number of legitimate births in France amounted to 1,817,572. Of these, 939,641 were male; and 877,931 were female. During the same years, the number of illegitimate births amounted to 140,566. Of these, 71,661 were male; and 68,905 were female. Amongst the legitimate births, the proportion of males is as 51,697 to 100,000; while amongst the illegitimate births, the proportion is only as 50,980 to 100,000. Now, the difference in the foregoing results might have amounted to 391 in 100,000 births, without surpassing the limits, within which the law may oscillate: but the actual difference very much exceeds this, and amounts to 717 in 100,000 births. This result shows, that some important difference exists in the *sum of the possible causes* of the two series of facts; and this difference really consists in the fact, that the births constituting one series [156] were in wedlock, and those constituting the other, out of it.[b]

The law of average number of children, born to each family, is to be ascertained by the same methods, and is subject to the same conditions. The sum of the phenomena and relationships, upon which, in each single instance, this number depends, is uncertain and variable; but the degree of this variableness is strictly confined within appreciable limits; so that the individual facts, although not *identical* with each other, are still *comparable* with each other. When a very large number of single instances have been accumulated, the average number to each ascertained, and the limits within which this number may oscillate measured, by an application of the calculation of probabilities, the law of which I am speaking is determined. But this law is uniform and permanent on condition, that the *sum of possible causes* of the number of children to each family remains the same. This sum may be materially affected by changes in the physical, the political, the moral, and the social condition of the people; and in this way the law itself, which is only the aggregate expression, or the generalization, of this sum, will also be affected. It follows, of course, that this law may vary at different periods of time, and amongst different people.

The foregoing doctrines are just as applicable [157] to many of the phenomena and relationships of pathology, as they are to those of physiology. Now, the laws, or principles, of pathology, of etiology, and of therapeutics, are ascertainable by the same methods, and subject to the same conditions. Each series of facts, or relationships, constituting the elements of the law, although not absolutely identical with each other, must still be sufficiently fixed and determinate in their character, to render them comparable facts; each series must consist of large numbers; and the limits, within which the observed average may oscillate, must be ascertained by an application of the calculation of probabilities; and the sum of possible causes must continue uniform. The law, whatever it is, — whether physiological, pathological, etiological, or therapeutical, — will be positive and absolute — the limits within which it may oscillate will become smaller — just in proportion to the degree of comparableness, or similarity, of the individual facts, the greatness of their number, and the fixedness, or uniformity, of the sum of their possible causes.

It is important, however, to observe, that there is a wide difference, in the readiness, facility, and positiveness, with which different laws may be determined. The aggregate of appreciable lesions, for instance, furnishing one of the elements in a group of morbid phenomena constituting a given disease, may often be ascertained with great certainty from a comparatively small number of ob-[158]servations. In the same way, the diagnosis of many diseases is susceptible of a comparatively ready and positive solution and settlement. The reason of this difference is to be found in the fact, that the phenomena and relationships, constituting the last-mentioned series of facts, are simpler and fewer, than in the more difficult cases; they approach nearer to the character of physical phenomena and relationships.

---

[b] Principes Généraux de Statistique Médicale. Par Jules Gavarret. p. 93, 94.

Amongst these laws, there is no one of so much interest and importance, as that of the therapeutical relationships of disease; and there is no one, the determination of which requires a more rigorous adherence to the methods and conditions laid down in the foregoing pages. Medical science has no problem, the solution of which is at the same time a matter of so much difficulty, and so much importance, as that involved in these relationships. For these reasons, I shall enter into a somewhat more detailed exposition of the subject before us in its connexion with therapeutics, or the treatment of disease; for the materials of which exposition, I am almost entirely indebted to the admirable treatise of M. Gavarret, on *Medical Statistics*.

The first condition, in the establishment of any therapeutical principle, or law, is this — that the facts, or phenomena, the relationships of which are to be investigated, shall be sufficiently fixed and definite to be *comparable*. The elements of this condition are thus stated by M. Gavarret. The [159] subjects of the disease, whatever it is, which is to be studied, ought to be taken from the same locality, and from the same classes of population; and the hygienic circumstances surrounding these subjects, during the treatment of the disease, should also be the same. These precautions, it is easy to see, are necessary, in order to render the individual cases of disease *comparable*. If the cases are taken from localities, differing in any important circumstances from each other, and also from classes of the population, differing, in like manner, from each other, it is obvious enough, that, from these circumstances alone, such peculiarities may be impressed upon the different cases of the disease, coming from one class and locality, or from another, as entirely, or in great part, to destroy their comparable character. Let us suppose, for instance, that the typhus fever of Ireland is the disease, the therapeutical relations of which we wish to ascertain. Nothing can be clearer, than that the law of these relations might be found to be quite different in subjects belonging to the lower orders, and living in insalubrious situations, and in those belonging to the higher classes, and living in healthy situations. The average physiological condition of these two classes, resulting from their very different habits and modes of life, might be so widely dissimilar, as to give to their disease a wide dissimilarity.[4]

In the second place, the disease, to be studied, [160] should be susceptible of a clear and positive diagnosis. It should be distinctly and accurately distinguished — nosologically, or as a species — from all other diseases; and it should be readily separable into its several varieties, so far as these are strongly enough marked to be of any importance. The necessity of this condition is so obvious, and the reasons of this necessity have been so fully pointed out in another place,[5] that there is no occasion for insisting upon it any further here. I will only add one or two remarks from Gavarret. When the law that we are in search of is that of the effects of any given plan of treatment, upon any given disease, considered nosologically, or as a whole, *every case of the disease that presents itself*, should be taken into account, whatever may be its stage, its degree of severity, or its complications. There should be no selection of case. The object before us is to ascertain the law of relationship between a given disease, as an integral morbid species, and a certain mode of treatment; and of course the disease should be taken as it presents itself, in all its varieties of degree, of period, and of complication. Under these circumstances, and when this is our object, the conditions in regard to locality, the occupation, and social position of the subjects, and so on, are of course to be disregarded. But instead of wishing to determine the results of any give method of management upon any given disease, as a whole, all its possible varieties and [161] complications, we may wish to confine our investigations to these results, in regard to certain varieties, or forms, of the disease. The solution of this latter problem is indeed of much greater practical importance, than that of the former; and at the same time it includes the elements of the former. Observation has long ago established the fact, that different forms, or varieties, of the same nosological affection, often require to be managed by methods more widely different, than are required by many dissimilar nosological diseases. The practical value of most therapeutical rules will be found to depend upon the applicability to certain forms, or varieties, of disease. When the object before us it to ascertain the effects of treatment upon these several forms of the same disease, it is necessary, to the legitimacy of our conclusions, that the case, constituting these forms, should be arranged in their several categories, at the earliest possible period of time in their progress. Each individual case must be placed in its appropriate series, or sub-division, constituting the particular form, or variety, to which it belongs, as soon as its character can be determined.

In the third place, the method of treatment which is to be applied should be defined as distinctly and as clearly as possible; both in its fixed and its fluctuating elements. When the foregoing conditions are fulfilled — when the subjects of the disease to be studied are taken from the same [162] general localities, and from similar classes of the population, thus securing a general similarity in their physiological tendencies, and susceptibilities — when they are exposed to the same hygienic influences during the continuance of their treatment — when the disease, whatever it is, is clearly and positively distinguished from all other affections, and susceptible also of being divided into its several forms and varieties, depending upon its extent or severity, — the period at which it was subjected to treatment, — the age, and sex of the patient, or any other appreciable circumstances; — and when, finally, the method of treatment is itself distinctly marked out, and well defined, we have secured our *comparable facts*, the legitimate data, and the only legitimate data, for our subsequent operations. It is not pretended, that these individual facts — any two of them even — are *absolutely identical*. The physiological condition of each single subject of the disease may have some peculiarity; this condition may differ in some respects from that of every other individual in any given series of cases — the disease itself may not be, and probably will not be, absolutely the same in extent and severity, in any two cases, even of its most distinct and well defined variety; — and, finally, the method of treatment may be subject to certain modifications in its application to each single case; — but notwithstanding all this, the facts are still *comparable* facts. Their degree of [163] difference is *limited*; this degree never surpasses certain definite and appreciable boundaries.

The phenomena to be compared having been thus ascertained and determined, we apply to them the methods, which have already been described. The law of relationship between the group of morbid elements, on the one hand, and the particular method of treatment, on the other; or, in other words, the effects of the treatment upon the disease can result only from an examination and analysis of a *great number* of individual instances, and by an application to the average result, of the calculation of probabilities. The law, whatever it is, may be relied upon, as positive and absolute, just in proportion to the *fixed* and *uniform* character of the compared facts, and to the *greatness of their number*; and, on the other hand, the law, if such it can be called, will be valueless, just in proportion to the opposite conditions. A failure in any one of the conditions destroys, just so far as it goes, the value and the legitimacy of our conclusions.

It is not necessary to the purposes of this essay, that I should enter into a full exposition and development of the principles of statistics in their application to the different branches of medical science. It is only by the aid of these principles, legitimately applied, subject to the conditions already pointed out, that most of the laws of our science are susceptible of being rigorously determined. I shall conclude this portion of my sub-[164]ject with one or two illustrations, taken from the work of Gavarret, showing the necessity of an examination and analysis of *large numbers of cases*, in order to arrive at any safe or positive results in regard to the effects of any particular remedy, or mode of treatment, in any given disease; and the danger of receiving the average observed result of any given treatment, as the true expression of the law, in all cases where the number of instances is small.

Louis, in his researches on typhoid fever, cites one hundred and forty cases; fifty-five of which were fatal, and eighty-five of which were not fatal; the mean mortality being equal to 0.37143, — or, in general terms, to 37 in 100. Now, an application, to this result, of the calculation of probabilities shows, that this average mortality derived from so small a number of cases may fluctuate between the proportions of forty-nine, and twenty-six, to a hundred; so that in comparing any other method of treatment with that of Louis, the aggregate sum of the conditions, or circumstances, remaining the same, it is not to be taken as settled, or certain, that the method is better or worse than his, unless the difference in the result surpasses, or exceeds, these possible limits. Let us suppose, that five hundred cases of a given disease have been subjected to a given treatment, with the result of one hundred deaths and four hundred recoveries; and that the same number of cases of the same disease have been subjected to a different [165] treatment, with the result of one hundred and thirty deaths, and three hundred and seventy recoveries. In the first class the ratio of mortality is as 20,000 to 100,000; in the second class, this ratio is as 26,000 to 100,000; the difference between the two being 6000 in 100,000. An application to these numbers of the law of probabilities shows, that the limit of possible variation is equal to 7,508 in 100,000; so that, although the second method of treatment may be better than the first, the number of cases by which the two methods have been tested is not sufficient to demonstrate, positively and rigorously, the fact of its superiority. By extending this observation to twice the number of cases, the ratio of mortality in each class remaining the same, we have the following results. The limit of possible variation, ascertained by the calculation of probabilities, when applied to a thousand cases, instead of five hundred, sinks from 7,508 in 100,000 to 5,306 in 100,000 which is considerably less than the observed difference in the ratio of mortality, this being as 6000 in 100,000. The result in this case, owing simply to the increase in the number of cases from which it is derived, demonstrates, positively, the superiority of the second method of treatment over the first.

It has already been stated, that in certain departments of medical science, the phenomena and relationships, with which the departments are concerned, may be more readily and certainly gene-[166]ralized; and the laws, or principles, constituted by these generalizations, may be established by the study and analysis of a much smaller number of cases; and that the reason of this is to be found in the greater degree of fixedness and uniformity in the phenomena themselves. Thus the diagnostic characters of many diseases, — of small pox, of measles, of scarlet fever, of pleurisy, of pneumonia, of rheumatism, of tetanus, of epilepsy, and so on, — are so constant and uniform, — the limits of their variableness are so narrow, — that it requires comparatively only a small number of complete and accurate observations to settle them definitely, and to establish their laws. The same thing is true of the appreciable lesions of many disease, — of phthisis,[6] of true apoplexy,[7] of pleurisy, of pneumonia, of pericarditis, and of others. But when we come to apply the foregoing rigorous doctrines *to what are commonly called the laws, or principles*, of therapeutics, how will these laws come out of the trial? Subjected to the ordeal of these doctrines, what becomes of the great mass of medical testimony to the efficacy of medical treatment? In how many instances, and to what extent, have the fundamental conditions of the establishment of any therapeutical law been fulfilled? How far have the facts been really *comparable* facts? In how many series of observations, has the nosological diagnosis, even, been established beyond any reasonable doubt; and, what is still more important, how accurately and clearly have the *varieties* [167] or *forms* of the disease been arranged in their appropriate categories? And even where these and the other essential conditions have been fulfilled, in how many instances have the observations been extended to a number of cases sufficiently large, to determine, with any positiveness, the actual results of the treatment upon the mortality of the disease? Alas! my brethren, there can be but one answer to all these questions; and humiliating as that answer may be, it is much better to make it, to hear it, and to give heed to it, than voluntarily to shut our ears and our eyes, and still stumble on in the dark. What is the character of the great mass of medical observation, in regard to the treatment of diseases, recorded in books and in medical journals? Dr. A. gravely reports a series of cases of what he calls tubercular consumption, all cured by his new method. But not a syllable is said about any evidence of the actual existence of the disease in any of his cases, derived from its physical signs; it may be only a year or two since the commencement of his observations; and no information is furnished as to the number of cases which have terminated fatally under the same management. Dr. B., with the same gravity, and apparent honesty, boasts, that he has been remarkably successful in the cure of *scarlet fever*; because he has not lost one of eight or ten, or it may be twenty cases, or *about* this number, of the disease, that have fallen into his hands, during the last season. With great self-complacency, he [168] compares the wonderful results of his own skill, with those of a neighboring practitioner, who, *he has understood, — and he has no doubt of the fact,* — has lost all, or nearly all of the cases of the same disease, which have unfortunately come under his care. The idea of inquiring how far the two series, or classes, of cases have been *comparable*, never seems to have entered his mind. Not a word is said, about the form or variety of the disease, which either he, or his neighbor, has been treating: although, supposing the results to have been as he has stated them, the probability is, that his own cases belonged to the simple form of the disease, and those with which he compares them to the anginose, or malignant form. Dr. C. announces to the medical world, that for the last year and a half, perhaps for the last four or five years, even, he has been uniformly successful in his treatment of croup. He says not a syllable about the form of the disease in the cases which he has managed; he has not ascertained whether they were cases of true membranous, or non-membranous croup. He may not be aware, that there is any such difference in the forms of this disease. On a further investigation into the real state of the facts, it may be found, perhaps, that the number of cases, of which he had kept no positive record, but which he really thought was very considerable, after all only amounted to some eight, or ten, or a dozen; and that from amongst these, even, he had excluded one case, because the child had been [169] scrofulous and feeble ever since its birth; and one other, because he did not see the patient till a day or two after the first appearance of the disease; and still a third, because it had not been properly treated by the physician who first had the care of it; and, finally, it frequently comes up, at last, that one case, which he had treated from its commencement, had terminated fatally, but it had entirely escaped his recollection. This sketch of the general character of medical testimony as to the effects of treatment, in these diseases, and in many others, is neither exaggerated, nor falsely colored. I appeal to the experience of all close and philosophical observers, now living; and to the multitudinous records on the pages of medical books and journals, for the proof of its faithfulness and its accuracy.

There is one remark of some importance which ought to be made here; and containing, as it does, a partial qualification of one amongst the many difficult conditions, conformity to which is essential to the establishment of any therapeutical law, or to the settlement of the positive and comparative value of different methods of treatment, it is a matter of no little consolation, that we are justified in making it. This remark is, that the number of cases, necessary to the determination of the actual or relative value of these different methods of treatment, is much less in certain diseases than in others. This is especially true, wherever the diagnosis is positive; and where, at [170] the same time, the issue of the disease, either in recovery, or in death, has already been ascertained to be very uniform and constant. Traumatic tetanus, hydrophobia, tubercular consumption and membranous croup, for instance, under all modes of treatment, have, thus far, in an immense majority of instances, terminated fatally. In these, and in all analogous cases, a widely different result, derived from the application of a new method of treatment, even to a *limited number* of cases, might be sufficient to determine very positively, the superiority of the method. The extent of the difference here, notwithstanding the smallness of the numbers, may exceed the limits of possible error, or fluctuation. Thus the recovery, under the application of a new method of treatment, of ten cases out of twenty, of hydrophobia, or traumatic tetanus, would constitute very positive evidence of its advantages, when compared with any other known methods. So, the application of a new method to a disease, the common termination of which in recovery, under other methods, had already been ascertained, with widely different and unfavorable results, even in a small number of cases, would be sufficient to determine very conclusively, its inferiority to the other methods.

But the extent to which these qualifying remarks are applicable is not very great. In a large proportion of the serious diseases, to which the human body is subject, the issue of the disease, [171] either in death or recovery, is a matter of much greater contingency and doubt. The ratio of mortality ceases to be extreme, in either direction; and in proportion as this happens, does it become necessary to augment the number of observations, from the study of which, any therapeutical law is to be derived. It may very naturally be asked, what, if these thing are so — if this hard doctrine is sound — is the practitioner of medicine to do? Is he to fold his arms, and to wait, till those who have the means and the ability, have gone through with these long, laborious, delicate and difficult investigations, — requiring so much time, and toil, and coöperation, — and have ascertained, positively, the actual and relative value of different modes of treatment in all the important diseases, which he is daily called upon to manage? Is his present knowledge of the effects of his remedies without positiveness and without value; and because it has not been obtained precisely by the methods, and subject to the conditions, above stated, is it to be distrusted and thrown aside? Is he no longer to bleed in acute pleurisy, or to give calomel in syphilis, or opium in spasmodic colic, or quinine in intermittent fever, because the therapeutical laws, in all these cases, have not been duly established and authenticated, according to the formulæ of the foregoing doctrine? Such questions, I say, will very naturally suggest themselves; it is proper that they should be answered; and the answer is this. The foregoing rules of medi-[172]cal treatment, and most others like them, have been ascertained and established, so far as they are ascertained and established, by a series of observations *of such vast extent*, as to compensate in a good degree, for the absence of the other conditions. In regard to many of them, the testimony of observers, for successive ages, has been nearly unanimous and uniform. The good effects of bleeding in most cases of simple, acute inflammation of the lungs, the pleura, the peritoneum, the pia mater and other organs and tissues, are so constant, as to leave no room for doubt or uncertainty. And the same thing is true of most of the generally admitted rules, or methods, of practice. This kind of observation has been sufficient to establish, in a general manner, these therapeutical maxims. They rest upon the concurrent testimony of immense numbers of witnesses; they are the results of an almost indefinite number of observations. It is to be taken for granted, that if these generally admitted rules, growing out of this very extensive observation, had been false and imaginary, the sagacity and experience of this host of witnesses could not have failed to detect their falsity. These rules have been, in this matter ascertained with a sufficient degree of positiveness, to render them our most valuable guides, in the management of disease. Although in very many instances the diagnosis of the disease, or the diseases, in question, must have been equivocal or mistaken; — [173] although the circumstances in which the patients were placed, and their individual conditions, must have been exceedingly diverse, still the aggregate number of cases has been so enormous, as to neutralize, in a great degree, the effects of these elements of imperfection and error. But it ought still to be added, that even in these cases, it is only by a faithful adherence to the rules and methods, which have been described, that the *exact value* of the several remedies, or modes of treatment, can be ascertained. These generally-received maxims of therapeutics are all still subject to revision. It is only by subjecting them to the rigorous discipline of the doctrines of this chapter, that their value can be absolutely and positively determined; and the actual and relative positions which they ought to occupy, definitively assigned to them.[c]

I wish now, in concluding this chapter, once more to call the attention of my reader to the [174] remark which I made at its commencement, to wit; — that the constituent elements of a law, or principle, in the science of life, do not differ from those of a law, or principle, of physical science. In both instances, the law, or principle, whatever it may be, consists solely and exclusively in the generalization, more or less rigorous and absolute, of the phenomena and relationships to which the law refers. The law, or principle, is not a creation of the reason; it is not the product of any *à priori* processes of the mind; it does not consist in any intellectual *deduction*, as it is termed, from the phenomena, or their relationships; it does not consist in any explanation, or interpretation, of these phenomena, or their relationships; — it is not to be found in anything superadded to them, or interposed between them; — it is the simple expression of their generalization, and nothing else. It may be well enough, perhaps, to remark here, although the grounds upon which the remark is founded must be sufficiently obvious, that the positiveness with which these principles are thus susceptible of being ascertained, applies to the *principles themselves*, and not to the *individual phenomena and relationships*, by the aggregate of which, they are constituted. Each of these separate elements of the principle, whatever it may be, is, in its very nature, contingent and variable, and must for ever continue to be so; and no possible degree of absoluteness in the principle can ever deprive these elements of this character. [175] How great soever may be the accuracy with which the average duration of human life, under all conditions, and in all circumstances, may be determined; the duration of any individual life will still remain, as before, altogether uncertain and contingent. And the same thing is true of pathological phenomena, and therapeutical relationships. How definitively soever the laws of these phenomena and of these relationships may

---

[c] It may even be said, I think, that the school of observation, whose principles and methods, I have endeavored to vindicate, in the present chapter, denies too peremptorily, and with too little qualification, the value of *all* results which have not been obtained in conformity to its own rigorous processes. One of my medical friends, says to me, in a letter, — "Perhaps there is one point that I may venture to caution you upon, — may I do so? I have sometimes thought that Louis and some of his disciples were a little rough in their treatment of unproved opinions; and that they showed rather too much pleasure in demonstrating that anything which seemed particularly probable, was not true. But I do not believe you will fall into this ultraism of the rigorous school." The remarks in the text will save me from this imputation.

be settled; the individual instances, or elements, of which they are composed, must still continue fluctuating and indeterminate, always, however, within certain limits; — the positiveness of the law cannot apply to the individual instances. The exactness of our appreciation of these instances, and our ability to estimate their precise value and conditions, may be aided by an acquaintance with the law; but this appreciation and estimate must still depend mostly upon the extent and accuracy of our knowledge of the several elements, which unite to make up the individual instances themselves. Thus although the ascertained law of the ratio of mortality in a given disease, under given circumstances, may assist us in predicting the termination, in an individual case; still this prediction must depend, in a great degree, upon our knowledge of the fluctuating and variable elements of the case itself. No acquaintance, however perfect, with the laws of pathology and therapeutics, can ever remove, or in any degree diminish, the necessity of a thorough and discriminating study [176] and knowledge of the single instances which unite to make up the materials of the law. Our diagnosis, prognosis, and management of individual cases of disease must depend, not so much upon the laws with which the diseases are concerned, as upon an accurate knowledge of the individual cases themselves; so that no perfection, or absoluteness, of the law, can ever lessen the necessity and importance of sagacity, discrimination, and skill, on the part of the physician, in the practical application of his art.

A vague and indefinite notion seems to have been long and extensively entertained, that some great principle, like the fact of gravitation, is yet to be discovered in physiological science, leading to results as new and magnificent, as those that flowed from the discovery of that simplest and sublimest of all known relationships. Even Cuvier exclaims, "*Why may not Natural History one day have its Newton?*"[8] And Whewell says: — "The idea of the vital forces may gradually become so clear and definite, as to be available in science, and future generations may include, in their physiology, propositions elevated as far above the circulation of the blood, as the doctrine of universal gravitation goes beyond the explanation of the heavenly motions by epicycles."[d] If the philosophy of this essay is not altogether mistaken and erroneous, the fallacy of all such expectations [177] must be sufficiently obvious. I trust that Natural History, including physiology and all its relations, will yet have, not one Newton, but many. Medical science — one of the branches of Natural History — has already had, indeed, not one Newton only, but many; and it is to their labors, that it is indebted for its existence, and for the degree of perfection, which it has been enabled to reach. But not to the development of any *abstract idea* of the *vital forces*; not to the discovery of any single and novel *principle*, as it is termed, has it ever been indebted, or will it ever hereafter be indebted, for its advancement. The "elevated propositions," of which Whewell speaks, whether in strict physiology, pathology, therapeutics, or whatever section of the science of life, are to be reached, not by any of the means, or processes, to which he seems to allude, but by the methods, and subject to the conditions, which have

[d] Hist. Ind. Sci. vol. ii. p. 405.

been already stated. These are the Newtons of medical science — Hippocrates,[9] Haller,[10] Morgagni,[11] Sydenham,[12] Hunter,[13] Laennec,[14] Andral,[15] Louis,[16] Chomel,[17] Du Chatelet,[18] — and others, — their worthy compeers, — who, imbued with the same spirit, guided by the same principles, and steadfast in their allegiance to the same doctrines, have resisted the influences of a fascinating but false philosophy, and have worked faithfully and diligently in their only true vocation, — *the study and analysis of phenomena and their relationships*; and the Newtons of our science, who are yet to come, must work [178] in the same direction, and their labors will be crowned with similar, but still nobler, more positive, and more valuable results.[e] [179]

CHAPTER XII.

PROPOSITION FOURTH

MEDICAL DOCTRINES, AS THEY ARE CALLED, ARE, IN MOST INSTANCES, HYPOTHETICAL EXPLANATIONS, OR INTERPRETATIONS, MERELY, OF THE ASCERTAINED PHENOMENA, AND THEIR RELATIONSHIPS, OF MEDICAL SCIENCE. THESE EXPLANATIONS CONSIST OF CERTAIN OTHER ASSUMED AND UNASCERTAINED PHENOMENA AND RELATIONSHIPS. THEY DO NOT CONSTITUTE A

---

[e] I have devoted no separate chapter to a formal exposition of what has been called the "numerical" method of observation. The reason of this omission must be obvious to every reader of my book. The doctrines of the numerical method, in its full development and application, are simply the doctrines of the foregoing chapter. This method is no new thing. Its elements are as old as Hippocrates: and there is hardly an individual writer on practical medicine, of any authority or importance, from his period to our own — including those who have been most unsparing in their abuse of the method — who has not used it. Every man, in every age, who has stated *numerable* facts in anatomy, physiology, pathology, or therapeutics, in *specific numbers*, has made use of the *numerical* method. Every observer, who counted accurately his cases of disease, or any of the phenomena connected with these cases, and gave the result *in numbers*, instead of resorting to the more common and indefinite terms — a *small number*, or a *large number*, *frequently*, or *rarely* — so far made use of this method. Its application to the facts and relationships of medical science had long been becoming more general and extensive, before the full measure of its value was practically exhibited by Louis, and its true principles philosophically developed and demonstrated by Gavarret. Although very slowly and reluctantly admitted by British physicians, as a formal and systematic *method*, it is nevertheless true, that some of the most distinguished and worthy amongst them, had adopted and used it somewhat extensively, many years before the publication of the researches of Louis. It is sufficient for me to mention, here, the names of William Woolcombe and John Cheyne, two stars of as steady and bright a lustre as any in the galaxy of British medical observers. This method, notwithstanding the opposition which it has met with from those who claim to be preëminently the disciples and champions of Hippocratic and rational medicine, has been constantly, though slowly, advancing in estimation, and pushing its way to favor in the British islands.

LEGITIMATE ELEMENT OF MEDICAL SCIENCE. ALL MEDICAL SCIENCE IS ABSOLUTELY INDEPENDENT OF THESE EXPLANATIONS.

The nature and value of what are called *Medical Doctrines*. Universal prevalence of medical hypotheses. Their bad influences. Methodism. Cullen's theory of fever. Homœopathy: State of its principles. Standard by which they are to be tried. Evil effects of *Medical Doctrines* upon the minds of medical men, and upon the interests of medical science. Broussais: His *History of Chronic Inflammations*, and his *Examination of Medical Doctrines*. Sydenham. How far interpretation may be allowed.

I HOPE, that the chapter on the nature of hypotheses in physical science, and their relations to science itself, has prepared the reader, if any such preparation was necessary, for what I have not to say upon the same subject, in its connexion with the science of life. The doctrines, which were advanced in that chapter, are, if I am not mis-[180] taken, all of them for still stronger reasons, and with less qualification, applicable to all the departments of the science of life. The essential character of all hypotheses, — both physical and physiological, — is the same; the nature of their constituent elements is the same; their relations to the respective sciences, with which they are connected, are the same. In the science of life, as in physical science, they consist, exclusively, in explanations, or pretended explanations, of appreciable phenomena and relationships, through the assumption of other unknown and imaginary phenomena and relationships. The science of life, in all its departments, is wholly independent of these pretended explanations; they do not enter into it, as one of its elements — they are, in no degree, and in no sense, one of its constituents.

It is also true, farther than this, that theory, or hypothesis, has played a much wider and more prominent part in the science of life, than in physical science. It has followed the former, like its shadow, from its birth, in the early ages of the world, to the present time. Under all circumstances, amongst all nations, in every stage and phasis of human progress, under the reign of all philosophies, and all religions; in all times, and everywhere, within the range of civilization, has medical science been attended with its protean hosts of hypotheses. These hypotheses have pervaded and ruled the science, and, to a great extent, determined its character. It is true, also, that the [181] influences of these hypotheses upon medical science have been more inauspicious and malign, than the influence of hypotheses upon physical science. Their effects have been bad, and only bad. The praise of having guided our researches, of having suggested new courses and new methods of investigation, of having assisted us in the conception and comprehension of phenomena, and in the expression of our ideas concerning them, which has been given to physical hypotheses, does not belong to these. They have only rendered more obscure and difficult what was sufficiently so before their investigation; and they have ever impeded the progress of the science which they professed to promote. Not only so, but they have almost always acted injuriously upon the practical application of the science of medicine. They have often destroyed, or neutralized, its efficacy as an art for the relief of human suffering. They have done more than this, even; — they have, in many instances, converted the science from an instrument of good, to an engine of positive ill — a means of inflicting upon men the very evils, which its true objects and aim are to remove. And these observations are, to a very considerable extent, as true of the present, as they are of the past. Hypothesis, in medicine, still passes for science — the former still usurps the functions, and claims the prerogatives, of the latter.

After the full consideration, which was given to this subject in the chapter on the hypotheses of [182] physical science, it is unnecessary to repeat the general remarks, which were then made, in their bearing upon the science of life. It will be sufficient for my purpose, to refer to some few of the hypotheses themselves; and, in this way, to try the truth and soundness of the doctrines, which I have ventured to lay down. In physical science, the number of these leading hypotheses is small, and they are generally characterized by a great degree of beauty, simplicity, and what, in a certain qualified sense of the word, may be called verisimilitude. In the science of life, they are without number; their name is legion; and, in most instances, they are as remarkable for their ill-adjusted complexity, clumsiness and improbability, as the theories of physical science are for the opposite qualities.

These theories, or hypotheses, in the science of medicine, are generally dignified with the title of *doctrines*. Thus, we have what are called the *doctrines* of the vitalists,[1] and the organists;[2] the doctrines of the humoralists[3], and the solidists;[4] the chemical, and the mechanical, doctrines;[5] the doctrine of irritability;[6] the doctrine of contro-stimulism;[7] the Cullenian, the Brunonian and the Broussaisian doctrines;[8] the doctrines of homœopathy,[9] of hydropathy,[10] and so on, from the beginning to the end of the long and heterogeneous chapter. It is not my purpose to write a history of medical doctrines, or, in other words, of medical hypotheses, — for all these so called doctrines [183] are only *hypotheses,* — and I shall speak of them, only so far, as may be necessary to the illustration of my own views. It is hardly worth my while, and it would aid but little in the direct elucidation of my subject, to say much of the medical theories of the Greek philosophers, either before or after the time of Hippocrates. The medical theories of these philosophers generally constituted a part of their more comprehensive theories of the universe, and consisted of similar elements. Hippocrates himself held no general doctrine in regard to diseases, which can properly be called a theory; a circumstance which now constitutes one of the highest and most legitimate titles to the preëminent position, which he occupies.

One of the first medical doctrines, or hypotheses, which was formally stated, and fully developed, was that of the *methodists,*[11] as they are called; and it is in this doctrine, that we find one of the earliest manifestations of that tendency to *dualism,* in pathological theory, which has never ceased to show itself, from that time to the present. According to this doctrine, the whole body was made to consist of a porous tissue, through which, fluids were constantly passing; and all disease was made to consist in the *relaxed,* or the *constricted,* state of the pores. This was a simplification of the doctrine of Asclepiades, according to whose system, many diseases depended, not merely upon the state of the pores, but upon the changes, and [184] the various actions upon each other, of the molecules passing through them. The latter doctrine was a mixture of humoralism and solidism; the former was pure solidism. The state of the pores, throughout the whole body, was inferred from the state of the skin, and from that of the natural outlets of the body. When the pores of the skin, or these outlets, were relaxed, or open, giving issue to the fluids of the body, the disease was said to belong to the class designated by the term *laxum*; when these pores and outlets were closed or constricted, the disease was said to belong to the opposite class, designated by the term *strictum*; and when some of the pores, or outlets, were closed, while, at the same time, others were open, the disease was said to belong to the class, designated by the term *mixtum*. Such was the doctrine, or hypothesis, of the methodists; and their therapeutics flowed necessarily from it; being founded exclusively on the double *indication*, of removing the two opposite conditions of the pores. This, although one of the oldest medical doctrines, or hypotheses, is, so far as its essential character and elements are concerned, an exact prototype and representative of all its successors. In order to interpret, and account for, the appreciable phenomena and relationships of morbid actions, certain properties and conditions of the body, wholly unknown and imaginary, are *assumed*; then, these supposed properties and conditions, by a second [185] *assumption*, are said to be connected with certain obvious states of the skin, and the natural outlets of the body, and, through this connexion, susceptible of being ascertained; and, finally, by a process of *à priori* reasoning, the treatment of all diseases, thus ascertained, is made to consist in the removal of these assumed and imaginary conditions; the therapeutics of the methodists naturally, necessarily, and *rationally*, as it is called, flowing from their pathology. Such, I say, when analyzed, and reduced to its actual elements, is the character of all medical hypotheses. Some of these may be more ingenious, than others, — some it would be more proper to say, may be less absurd, and preposterous, and improbable, than others; but they are all essentially alike; they all consist in certain unknown and imaginary phenomena and relationships, assumed for the purpose, as is vainly supposed, of rendering more intelligible to our comprehension, of explaining, interpreting, and accounting for, the phenomena and relationships, which are obvious and appreciable. They constitute, in no sense, and in no degree, any legitimate element of the science of life.

It is curious to see, in this ancient and venerable hypothesis, some of those more strongly marked features, which have never ceased to reappear in the successive members of the prolific family to which it belongs. Thessalus,[12] like his modern disciples, and in strict keeping with the spurious but seductive simplicity of his pathological creed, said, [186] that he could make of the most illiterate artisans excellent practitioners in less than six months. Cœlius Aurelianus,[13] with all his merits, like other members of his sect, denied the existence of specifics, because their effects could not be attributed either to *constriction* or *relaxation*; and banished purgatives from his materia medica, because their action could not be referred to either of his two imaginary *modi operandi*. There are treatises on therapeutics, still fresh, both from British and American presses, imbued and pervaded by the same *à priori rationalism*.

Passing over the chemical, mechanical, and humoral doctrines, with their various modifications and combinations, let us come down nearer to our own times, and look at some one or two of the more recent pathological theories, and see if they have any better claim, than their predecessors, to be considered as anything more, than gratuitous conjectures or speculations. One of the most celebrated of these, — constructed with great care and skill, all its parts adjusted and arranged with a formal and elaborate exactness worthy of its famous author, — is the Cullenian theory of fever. This theory begun by *assuming*, that the cause of the cold stage of a febrile paroxysm is the cause of all the subsequent phenomena. The doctrine *assumed*, in the second place, that this primary cause is to be found in the weakened energy of the brain, occasioned by the application, and action upon it, of certain sedative influences, or agents. Then, it [187] was further *assumed*, that this diminished energy of the brain produces a state of debility in all the functions of the body, but especially in the heart and arteries, and in the extreme vessels; in consequence of which it was again *assumed*, that these vessels become the seat of spasm. In consequence of the cold stage, and of this spasm of the extreme vessels, it was finally *assumed*, that the heart and arteries are excited to increased activity, and by this activity, the spasm of the vessels is overcome, the energy of the brain is restored, and the series of morbid actions thus entirely destroyed. With all this, the *vix medicatrix naturæ* is so strangely mixed up, that it is not easy to get at the exact ideas of the author himself, in regard to its functions and agency. But such, at any rate, briefly stated, is Dr. Cullen's *doctrine of fever*. He seemed to think, that it was a very sound, a very philosophical, and a very *useful* doctrine. "I flatter myself," he says, "that I have avoided hypothesis, and what have been called theories!" Now, I have no intention of entering into any examination of the *doctrine*, as its author calls it, or of indulging in any comments upon it. I cite it only as an illustration of the doctrine of this chapter. Certainly, the wildest dreamer in pathology, and the loosest *à priori* reasoner, even, could hardly have gathered together a jumble of *assumptions*, more utterly gratuitous. They are as improbable, each in itself, as they are altogether incoherent and heterogeneous. But the entire the-[188]ory differs, in no way, so far as its essential character, and its relations to true science, are concerned, from those of the methodists, the chemists, the mechanicians, amongst the ancients; or from that of Brown, of Rasori, of Broussais, of Hahnemann, or of Samuel Thompson, amongst the moderns.

I have spoken of Hahnemann; and I will conclude this kind of illustration, by a short examination of what is called the homœopathic system of medicine. It is possible, perhaps, that some of my readers may be surprized, that I should thus recognize the claims, or pretensions, of this system to the character of a *medical doctrine*. But its claims are just as legitimate, as those of any of the systems, of which I have already spoken. They are of the same nature; they rest upon the same grounds; they differ, in no respect, from the claims of Methodism, Cullenism, Brownism, or Broussaisism. Whether there has been, or has not been, more charlatanry amongst those of other doctrines, it in no way concerns my present purpose to inquire. The system, I have said, claims our suffrages, on the same grounds, that are set forth by all other systems; and I intend to test its soundness by an application to it of the same philosophical principles, by which those other systems have been tried. This, certainly, its friends and advocates cannot complain of. I will not condemn it, on the ground of any apparent im-[189]probabilities, or absurdities, which it may involve. I am ready to admit and to believe any and all of *its* assertions, on the same conditions, upon which I admit and believe any and all other assertions. I shall not endeavor to ridicule its infinitesimal doses, nor that element, in its pathology, which refers hysteria, mania, epilepsy, every species of spasm, softening of the bones, cancer, fungus hæmatodes, gout, hæmorrhoids, dropsy, epistaxis, hæmoptysis, asthma, suppuration of the lungs, impotence and sterility, deafness, cataract, gravel, paralysis, all kinds of pains; and very many other chronic diseases, besides a large majority of acute diseases, to *psora*, or *itch*, as their only true, fundamental and productive cause! All this, and much more, even the assertion, that a homœopathic dose of mesmerism will snatch from impending death a case of uterine hemorrhage,[a] I am quite ready to admit and receive, as true and sound doctrine, whenever it is so established, according to the philosophy of this essay, — but not till then.

The leading principles of the homœopathic doctrine may be thus stated. I derive them from the French translation of Dr. Hahnemann's exposition; and whatever modification they may have undergone, in the hands of his successors, can in no way affect their relations to the true philosophy of medical science. I may say the same thing of [190] the details of these principles; these details have no bearing upon my present purpose.

1. To the entire human organization, is superadded an immaterial principle, — a dynamical, or moving, force, — active by itself, — by which, the organization is ruled and controlled. It is this dynamical force, or principle, upon which all morbific causes or influences act; and the disturbance, which these causes occasion in this principle, operates of necessity upon the organization, deranging its healthy actions, and perverting its natural sensations.

2. Every modification of this immaterial and independent principle, through the altered actions and deranged sensations of the organs, which it governs and moves, manifests itself by external signs, or symptoms, which are always recognizable and appreciable, by the attentive and careful observer; so that *the totality of the symptoms*, in any given case, becomes an absolute and infallible index and exponent of the changes in the organs, or, in other words, of the disease. These changes, themselves, are beyond the reach of our investigation, so that the study of anatomical lesions is only a vain dream.

3. The vital force being a dynamic power, the morbific causes, occasioning its disturbance, can do this only in virtue of a like dynamic power in themselves; and these disturbances, thus produced, can be removed only by modifiers, or remedies, equally dynamic in their character, and acting on the vital force. [191]

4. The effects of all modifiers, or remedies, upon this force can be certainly and positively ascertained, only when the force itself is not already disturbed by the action of morbific causes, — or, in other words, — when the body is in a state of perfect health. The action of these modifiers is constant and uniform; so that when

[a] Exp. de la Doc. Homœop. p. 292.

they act as remedies it can only be by modifying the vital force precisely as they do in health.

5. The totality of the symptoms, and the disease, being, so far as our knowledge is concerned, equivalent terms, or the same thing; the former being removed, it follows, of necessity, that the latter is cured.

6. This cure can be accomplished only in two ways, — first, by exciting, through the agency of modifiers, or remedies, actions in the vital force *like* those which already constitute the disease; or, second, by exciting actions in this force *unlike*, or opposite, to those constituting the disease.

7. All pure experience, and all careful trials, show that the latter is impossible; and that even when the symptoms are diminished, or removed, by it, they never fail to reappear in an aggravated form. It follows, then, that there is only one method by which the totality of the symptoms, representing the disease, can be certainly and permanently removed; and that is through the agency of those substances and influences, which so modify the dynamic force of the healthy body, as to produce [192] a totality of symptoms *like* those which represent the disease.

8. The artificial action, constituting this totality, must be a little stronger, or more powerful, than that representing the disease.

9. Pure experience shows, that all true remedies do act in this manner; and do cure diseases. All opposite, or allopathic experience, as it is called, is false and deceptive. Diseases are never removed by substances, which do not act in this manner.

10. Remedies, or modifiers, in order to produce the desired effect on the disturbed vital force, must be introduced into the body in exceedingly minute, and almost infinitesimal quantities.

Such I believe to be the fundamental principles of the homœopathic doctrine. I have endeavored to state them as clearly, and explicitly, as possible. Their details, their practical application, their illustrations, and the reasoning by which they are supposed to be supported, do not at present concern us. My single purpose is to see how far they are conformable to the philosophy, which it is the design of this essay to vindicate and establish. Are these *principles*, as they are called, true principles, according to the legitimate and philosophical meaning, which ought to be attached to this word? Do they consist of phenomena and relationships, of an appreciable [193] and positive character, ascertained by absolute and extensive observation? Let us see.

How is it with the first, fundamental proposition, upon which all the others are made to depend, and from which they flow? What is the *material* of which this foundation consists; upon which the entire homœopathic superstructure is made to rest? Is this proposition, *fact*, or *fancy*? Is this foundation wrought from the adamant of positive phenomena, or is it woven with the tissue of dreams? It is not possible that there can be but one answer to these questions, unless the answer comes from a dreamer. *There is no evidence, whatever, of the existence, even* — to say nothing of its alleged properties and relations — *of this independent, dynamical force, presiding over, and moving, the organic structure*. The existence of this force is an *assumption*, just as perfectly and entirely gratuitous, as it is possible to imagine. It is more so than that of the *strictum* and *laxum* of the methodists, or the *spasm* of Cullen. The whole doctrine of this dynamical force is nothing but physiological transcendentalism. *Life is the sum of the organization, and its actions. This is all we know — this is all we can know, about it. What* the vital force is — *how* it is connected with the organic structure — the nature of the bond between them — the intimate manner in which each is acted on by its modifiers — is utterly unknown to us; and the probability is, that this ignorance will never be removed. This ele-[194]ment in the doctrine of Hahnemann is no new thing; it is very much like the *archeus* of Van Helmont,[14] and other old systematists, and the evidence of its existence is of just the same character.

In regard to the second proposition, it is not enough to say that it is gratuitous; it is worse than this. It is in direct and unqualified opposition to the most extensive and positive observation. It is not true, that every modification of the condition of the living structure and powers has its invariable and characteristic external sign, or manifestation, through which the modification is made known. Certainly, it is by their signs and symptoms, that internal diseases are revealed to the physician. But daily observation shows, that there is no uniform and invariable relationship between the extent and intensity of the disease, and its external signs. The prominency, the number, and the combination, of these, depend upon many circumstances beside the disease with which they are connected. Has no change taken place in the condition of the living structure, or its actions — in the relations, susceptibilities, and tendencies, of one or both — during the latent period of the contagion of small-pox, yellow fever, or hydrophobia? Is this independent vital force of homœopathy — admitting it to be present, with all its assumed properties — in no way affected by this poison of terrific energy, that has crept into the system? It is impossible [195] to suppose, that such can be the case; but the modification, whatever it may be, gives no outward and intelligible sign of its existence. Neither is it true, as is alleged by homœopathy, in connexion with this subject, that the internal changes in the organs themselves, are wholly beyond our means of investigation. To a very great extent, they are entirely *within* our means of investigation, and they constitute one of the most valuable and positive elements in our knowledge of disease.

The third proposition, asserting the existence of certain properties and susceptibilities of the dynamic vital force, is like the first, in regard to the separate and independent existence of the force itself, wholly gratuitous.

The fourth principle in the doctrine of homœopathy is, that the remedial action of all substances can be ascertained only by the effects which they produce upon the dynamic power, in its undisturbed state. The doctrine, that all therapeutical laws consist in ascertained relationships between morbid conditions, on the one hand, and their modifiers, on the other, has been so fully stated, that it is unnecessary to say anything further upon this opposite principle of Hahnemann. I will not comment in detail upon the several other propositions, as I have arranged and numbered them. The principles which they profess to set forth are not *principles*, but *assumptions*. There is no proof, that diseases can be removed [196] in only two ways, or in only one of two ways. There is no proof, that remedies act on the assumed vital force, by producing a modification *like* that in which the disease consists. We have no knowledge, whatever, of the intimate and ultimate action of modifiers, or remedies, on the structure or susceptibilities of the body. All these elements of the doctrine before us, so gravely set forth as facts, are anything but facts. They are all "such stuff as dreams are made of," and nothing else. The whole system of Hahnemann, from beginning to end — in its principles, and in its details — is one of unadulterated and arrogant dogmatism, resting exclusively upon *à priori* reasoning, or, in other words, upon mere speculation.

But, it will probably be said, — the doctrine is sustained by facts; its soundness and correctness are corroborated and demonstrated by the results of observation, — it professes to rest upon experience, as well as upon reason, and the nature of things. The experience upon which Hahnemann founds his doctrine, and by which he professes to sustain it is, if this is possible, more fallacious, and less philosophical, than the doctrine itself. I only insist, that this experience shall be tried by the same test, as has been applied, in this essay, to *all* medical experience. Let my readers examine the experience which Hahnemann calls in to support his doctrine, and refer it to the rules, which have already been laid down, as applicable [197] to all experience in medical science. His work is full of bold and unqualified assertions upon this subject, I admit; but the evidence of the experience itself is utterly wanting. In the entire history of medical doctrines, there is not one in regard to which the proof of their soundness derived from experience is so entirely defective and unsatisfactory, as it is here. Perhaps the most striking fact running through the whole exposition, or Organon, of Hahnemann, is the *absolute nullity* of all conclusive observation. He says, with no qualification, whatever, "the allopathic method never really cures;" — "the homœopathic method never fails to cure;" but when we look for any evidence of the truth either of one allegation, or the other, it is nowhere to be found. There is no evidence, in the first place, at all conclusive, of the power of the remedies themselves, to produce, in the healthy body, the effects that are so confidently attributed to them. The author of the system lays down a general law, which he wishes us to regard as invariable and absolute, — for instance, that similar diseases must and do cure each other, — the stronger disease always curing the weaker, — and then he gives such facts as the following to prove it. Small pox is often complicated with opthalmia and dysentery, — they are similar diseases. Dezoteux and Leroy report each a case of chronic opthalmia, cured by inoculation; and the occurrence of small pox cured a dysentery in a case reported by Wendt. Then [198] another law, — equally universal and absolute, — is established by the following evidence, and by a few other similar cases: — Tulpius tells us, that two children, having contracted *tinea*, were free from attacks of epilepsy, to which they had been subject, so long as the tinea continued.[b] The worthlessness of all such

[b] The value attached by Hahnemann to simple experience is very unequivocally manifested by a direct admission in his Organon, of the subordination of its authority to that of his *rational*, or *à priori* principles. He says, that the true physician will be cautious how he suffers himself to become attached to any particular remedies, *merely because he has often employed them with success*; and that he will, in like manner, also, be cautious how he suffers himself to be prejudiced against remedies, for the opposite

experience has been fully shown in another part of my essay; and it is upon such experience, that homœopathy, — apart from its *à priori* doctrines, — urges its claims to our consideration. The efficacy, and advantages, of its mode of treating disease, can be established in only one way, — by only one method. Let it produce its comparable facts, — its cases of disease, clearly distinguished, and separated from other diseases, both in their nosological diagnosis and in their varieties, — and let it produce these in large numbers, — not in groups of twos or threes, or of twenties and fifties, even, but of hundreds; — let it conform to the rigorous and indispensable conditions, which have already been so fully stated, [199] let it follow the methods, which have already been laid down; and then, and not till then, true philosophy will give heed to its words. These things it has not yet done; it cites in its favor only the loose tongue of common report; the same tongue that proclaims, *with like confidence, and on precisely similar proof*, the superiority of Swaim's panacea, Brandreth's pills,[15] and lobelia.[16] The smallness of the homœopathic doses; its apparently improbable and exclusive mode of treatment, constitutes no philosophical and valid objection to the system itself. I do not deny its claims, on these grounds, or for any such reasons. I doubt and deny them, — so far I mean as the results of its practice are concerned, — solely because they are not established by competent observation. Whenever they can be so established; whenever, in conformity to the conditions of all conclusive and satisfactory experience in therapeutics, it is shown, that homœopathic treatment cures diseases with more readiness, ease and certainty, than other treatment does, I will at once embrace and believe it; in no way prevented, or influenced, by any *à priori*[17] considerations against it whatever.

Its fate as a *doctrine* is certain and inevitable. Of this, the voices of all medical history are here to inform us. Like all its forerunners, and like all its conceivable successors, constructed on false principles; consisting, not of positive facts and their relationships, clearly ascertained, and suitably classified; but of gratuitous assumptions of facts [200] and relationships, altogether imaginary, one only possible doom awaits it. After living its short day of sunshine in the popular and professional favor, it will follow in the footsteps of its departed predecessors, — methodism, chemicalism, humoralism, mechanicalism, Cullenism, Brownism, Broussaisism, Rushism,[18] Cookism,[19] Gallupism,[20] and all the host of other rational *isms*: —

> "It shall be borne to that same ancient vault,
> Where all the kindred of the Capulets lie; —"

there to rest, as in the spirit and inspiration of a better philosophy, we may not undevoutly hope, in a sleep that shall know no awakening; its final departure rendered somewhat more respectable perhaps, although hardly accelerated, by the

---

reason, that they sometimes fail to succeed. He must never, he says, lose sight of the grand truth, that amongst all remedies, one alone merits the preference, — that which produces symptoms nearest like those characterizing the disease for which it is to be given. *Organon*, p. 271.

glittering arrows from the full quiver of Holmes, which are trembling in its heart.[c] [201]

---

[c] To attempt any philosophical analysis, or to enter into any general consideration, of the causes of the rapid diffusion through the popular mind, and of their strong hold upon it, of medical delusions, does not fall within the scope of this essay. My readers will all thank me, however, for the gratification, which I indulge, myself, and furnish to them, by the following extract from a late address, by one of my former colleagues in medical instruction — a gentleman as remarkable for the sparkling brilliancy of his imagination, as for the extent of his medical attainments, and the soundness of his medical philosophy. "Society is congratulating itself," he says, "in all its orations and its periodicals, that the spirit of inquiry has become universal, and will not be repressed; that all things are summoned before its tribunal for judgment. No authority is allowed to pass current, no opinion to remain unassailed, no profession to be the best judge of its own men and doctrines. The ultra-radical version of the axiom, that all men are born free and equal, which says, 'I am as good as you are,' and means, 'I am a little better,' has invaded the regions of science. The dogmas of the learned have lost their usurped authority, but the dogmas of the ignorant rise in luxuriant and ever-renewing growths, to take their place. The conceit of philosophy, which at least knew something of its subjects, has found its substitute in the conceit of the sterile hybrids, who question all they choose to doubt, in their capacity of levellers, and believe all that strikes their fancy, in their character of reverential mystics. This is the spirit which you will daily meet with, applied to your own profession, and which might condense its whole length and breadth into the following formula: A question, involving the health and lives of mankind, has been investigated by many generations of men, prepared by deep study and long experience, in trials that have lasted for years, and in thousands upon thousands of cases; the collected results of their investigations are within my reach; I, who have neither sought after, reflected upon, nor tested these results, declare them false and dangerous, and zealously maintain and publish, that a certain new method, which I have seen employed once, twice, or several times, in a disease, of the ordinary history, progress, duration, and fatality, of which I am profoundly ignorant, with a success which I (not knowing anything about the matter,) affirm to be truly surprising, is to be substituted for the arrogant notions of a set of obsolete dogmatists, heretofore received as medical authorities.

"What difference does it make, whether the speaker is the apostle of Thomsonism, the 'common sense' scientific radicalism of the barn-yard, or homœopathy, the mystical scientific radicalism of the drawing room? It is the same spirit of ignorant and saucy presumption, with a fractional difference in grammar, and elegance of expression. If this is just, it affords you a hint as to the true manner of dealing with such adversaries. Do not think that the special error they utter before you, is all that you have to vanquish. *The splinter of stone at your feet, which you would demolish with your logical hammer, runs deeper under the soil of society than you may, at first, imagine; it is only the edge of the stratum, that stretches into the heart of the blue mountains, in the far horizon.* Think not to gain anything by arguing against those who are drunken upon the alcohol, hot from the still of brainless philanthropists; who are raving with the nitrous oxide, fresh from the retort of gaseous reformers. Argument must have a point of resistance, in a fixed reasoning principle, as the lever must have its counter-pressure in the fulcrum; no mariner would hope to take an observation by an ignis fatuus, to steer by a light-house, floating unanchored upon the tempestuous ocean! No, your object must not be this, or that, heretical opinion, but the false philosophy, or the shattered intellectual organization from which it springs; it is Folly who is masking under the liberty cap of Free Inquiry; it is Insanity who has wandered from the hospital, without his keeper!

"After what I have just said, you cannot think that I shall waste your time, with allusions to the particular vanities that happen to engross the medical amateurs of our community, at this precise moment. On some occasions, and before some audiences, it may be justifiable, and perhaps useful, to show up some extreme and insupportable extravagance, as an example, not for the sake of the sharpers, who live by it, or the simpletons, whom they live upon, but for that of a few sensible listeners, who are disturbed by their clamor, and wish to know its meaning. Even then you must expect a shoal of pamphlets to spring upon you, with the eagerness of sharks, and the ability of barnacles. You have given a meal to your hungry enemies, by merely showing yourself, like an animal that ventures into a meadow, during the short empire of the horse-flies.

I have said that pathological theories, or hypotheses, like those in the physical sciences, have generally been framed for the purpose of explaining, or interpreting,

---

"I know too well the character of these assailants, to gratify their demand for publicity, by throwing a stone into any of their nests. They welcome every cuff of criticism, as a gratuitous advertisement; they grow turgid with delight, upon every eminence of exposure which enables them to climb up where they can be seen. Little as they know of anything, they understand the hydrostatic paradox of controversy; that is raises the meanest disputant to a seeming level with his antagonist; that the calibre of a pipe-stem is as good as that of a water spout, when two columns are balanced against each other. They would be but too happy to figure again in the eyes of that fraction of the public, which knows enough to keep out of fire and water, and to quote that famous line from the idiot's copy book,

'Who shall decide, when doctors disagree?'

"As I have given them more prose than they are worth, allow me to toss them a few lines, written for a recent anniversary, which, if they are unworthy of your approbation, are quite good enough for them.

"The feeble seabirds, blinded in the storms,
On some tall light-house dash their little forms;
And the rude granite scatters for their pains,
Those small deposits which were meant for brains.
Yet the proud fabric, in the morning sun,
Stands all unconscious of the mischief done;
Still the red beacon pours its evening rays,
For the lost pilot, with as broad a blaze;
Nay, shines all radiance o'er the scattered fleet
Of gulls and boobies, brainless at its feet.

I tell their fate, but courtesy disclaims
To call our kind by such ungentle names;
Yet if your rashness bid you vainly dare,
Think on their doom, ye simple, and beware.

See where aloft its hoary forehead rears,
The towering pride of twice a thousand years!
Far, far below the vast, incumbent pile,
Sleeps the broad rock from art's Ægean isle;
Its massive courses, circling as they rise,
Swell from the waves, and mingle with the skies;
There every quarry lends its marble spoil,
And clustering ages blend their common toil;
The Greek, the Roman, reared its mighty walls,
The silent Arab arched its mystic halls;
In that fair niche, by countless billows laved,
Trace the deep lines that Sydenham engraved;
On yon broad front, that breasts the changing swell,
Mark where the ponderous sledge of Hunter fell;
By that square buttress look where Louis stands,
The stone yet warm from his uplifted hands;
And say, O Science, shall thy life-blood freeze,
When fluttering folly flaps on walls like these?"

— *The Position and Prospects of the Medical Student. An Address delivered before the Boylston Medical Society of Harvard University, January 12, 1844.* By OLIVER W. HOLMES, M. D.

the appreciable phenomena [202] and relationships of disease. But it is important to remark, that in many instances they are not even entitled to this credit; for they seem to have been the spontaneous product of that tendency in [203] the mind to wild and fanciful speculation, which can be held in abeyance only by the stern discipline of positive ideas, and a sound philosophy. Sometimes they have probably grown out of a [204] vague notion, that by their adoption, and through their agency, the science of medicine could be made to approach nearer, in some respects, to the simplicity, certainty, and absoluteness of physical science. These *à priori* abstractions, under the misnomer of laws, or principles, were supposed, it would seem, to take the place, and to perform the functions, of laws, or principles, in physical science. It is difficult to account for the importance attached to them by their authors, on any other grounds. They explained nothing, in any intelligible sense of the term; they interpreted nothing; they accounted for nothing.

I trust, that the true character of all these pretended medical doctrines is now sufficiently obvious to the reader. I hope he is prepared to judge them according to their deserts, and to assign them their appropriate position *without* the pale of legitimate science. But before leaving this subject, I wish to make a few remarks upon the evil influences which they have exerted, and which they still continue to exert, upon the minds of those who have faith in them; and upon the progress of medical science.

The art of observation is always a very difficult [205] art; and nowhere is it more so, than in the science of medicine. It is one of the rarest accomplishments; and although the annals of medical science are crowded with the names of men, who were famous for their learning, or for their reasoning and speculative powers, they bear those of but few who were distinguished as observers. We have hosts of erudite and ingenious builders of systems, but only one Hippocrates and one Sydenham. A good observer in medicine must be furnished with quick and accurate senses; and his mind, besides being clear and comprehensive, must be free from all scientific prejudice, bias, or passion. Then, he must be educated to the art of observation; both his senses and his mind must be trained, by a long course of appropriate discipline and practice, before he can become skilful and accomplished in his calling. It is well known, that it was not till after many months of assiduous labor, in the business itself of observing, that Louis found himself at all prepared for the task, which he had undertaken, of studying anew, and more carefully than had been done before, the phenomena and relationships of disease. Now, one of the first and most inevitable effects of a belief in any *à priori* system of medicine is an *utter disqualification of the mind for correct and trustworthy observation.* No man with one of these hypothetical crotchets in his brain is to be trusted. Every object about him is discolored and distorted by this doctrinal medium through which he sees it. His intellect-[206]ual vision is neither true nor achromatic. He will always find what he expects to find; and he will always fail to discover what he has concluded beforehand will not be present. And this may be said without impugning his good faith, and his honesty: although it can hardly be regarded as uncharitable, to assert, that the mind must be strongly armed with integrity, and singularly free from the infirmities of human nature, to escape, wholly, worse effects, than those that I have spoken of, growing out of a blind adherence to any of these systems, and the controversies with which they are always attended. John Brown would have found unequivocal signs of debility, while Dr. Clutterbuck[21] would have discovered an inflammation of the brain, Broussais a gastro-enteritis, and Dr. Cooke a congestion of the veins of the liver. I have just mentioned Broussais; and I may add, that nowhere can a more striking exemplification of the influence of which I am speaking be found, than in the history of his mind. His two great works are the *History of Chronic Inflammations*, and the *Examination of Medical Doctrines*. The former is almost entirely a work of pure observation. It was written while his mind was yet free from the narrowing, darkening, and distorting influences of a blind faith in a doctrinal, medical creed. Considering the time at which it was written, and the circumstances under which its materials were gathered, it may justly be regarded as one of the most remarkable works of [207] practical medicine in any language. It opens with this sentence: — "*La medecine ne s'enrichit que par les faits*:" — *Medicine is enriched only by facts*; and the spirit of these words runs through and presides over the whole work. The clear-headed, sagacious, and discriminating observer shines out in every one of its pages; they are all luminous with practical wisdom. I am sure that no man, at all capable of appreciating it, can read this book, especially the first parts of it, treating of diseases of the lungs and pleura, without feeling, that it proceeded from a mind of extraordinary capacity and strength; and without entire reliance on the accuracy and good faith of the author, as an observer. What precision and positiveness in his diagnosis! What enlarged but cautious comprehensiveness in his general conclusions! What honesty and frankness in his admission of the frequent impotency of medical art! What admirable tact and discrimination in his selection and use of remedial measures! How clear and sound the philosophy, which illuminates and binds all this together! How true his appreciation of the emptiness and worthlessness of theoretical speculations; equal almost to that of Newton and Davy! Nowhere in his pages does the doctrinal partisan show himself; or if at times that yet undeveloped tendency of his mind, which afterwards transformed him from the calm, dispassionate, and philosophical observer, into the fierce and excited head and leader of a sect, indicates its presence, [208] and shows something of its latent activity, it is held in strict subordination to his better judgment. It is never suffered to usurp dominion over the definite convictions, and positive ideas, resulting from the simple study and analysis of the phenomena and relationships of disease.

But in his *Examination of Medical Doctrines*, all this is far otherwise. Broussais had now become an *à priori* medical philosopher; he had framed a creed of *rationalism*; he had established a new doctrine of his own; he was the acknowledged chief of a new party; a single dominant idea had taken possession of his mind. In this work, as in the other, the traces of his great genius are still evident. The now "*bad eminence*" of his strong intellect still shines through its pages. His rapid and vigorous thoughts still clothe themselves in his sturdy and glowing phraseology. His greatness he could not put off, if he would; but the scientific rectitude of his mind is no longer present; the clearness of his vision has become obscured; the acute and circumspect observer of diseases, and their relationships, indifferent as to the result of his investigations, provided only, that this result was the expression of the actual truth, is now the interested seeker for certain particular phenomena, *which he wishes to find*; the upright and impartial judge has become the *ex parte* advocate and witness. And, as generally happens in similar circumstances, not only is his mind perverted by the influences of a false philosophy, but [209] his passions are excited by the controversies which grow out of it. His arrogance and dogmatism are as offensive as his criticisms of those who refuse to follow him are injurious and unjust. The exigencies of his own creed led him into inconsistencies, and his contradictions of himself are as direct and flagrant, as they are humiliating.[d] He has himself become an illustration of the reasonableness and propriety of one of his own sayings: — "I hold it as a principle always to suspect the experience of a man whose mind is preoccupied." He remarks of Lord Bacon, that he often sacrificed at the altar of one of the idols which he had overthrown. And he too, it may be more truly said, redoubtable iconoclast as he is, has set up as false an idol, as any which he has broken; and declared a vindictive and uncompromising warfare against all who refuse to fall down before it. And such are the natural, and almost the inevitable, results of a belief in any of those *à priori* systems. [210]

But this is not all. A belief in these doctrines not only disqualifies him who holds it, *as an observer of disease*; it unfits him, to a greater or less extent, *for the practice of his art*. The builder up of an artificial and *à priori* system of pathology necessarily *deduces* from it a corresponding *à priori* system of practice. His faith in the latter is just as blind and implicit, as his faith in the former. All medical history confirms the truth of this remark. The therapeutics of the systematist is always deduced from his pathology, and rests, of course, upon indications as imaginary and hypothetical as the pathology itself. The only legitimate indication of the methodist was to remove the *laxum*, and the *strictum*, which constituted his diseases; and his means were chosen, with reference to their supposed fitness for this purpose. The chemical pathologist, who made all disease to consist in the preponderance of an acid, or an alkali, must fulfil the only rational or possible indication, that could present itself to his mind, by the administration of an appropriate acid, or alkaline neutralizer. John Brown, for the removal of [211] his hypothetical asthenia, must

---

[d] One of the most flagrant instances of this unblushing self-contradiction, and inconsistency, occurs in connexion with Broussais's remarks on the work of Prost, published in 1804. In the first edition of his *History of Chronic Inflammations*, after citing the opinion of Prost, on the agency of inflammation of the digestive mucous membrane in the production of ataxic fever, he says, — "I have too often found this membrane in good condition after the most malignant typhus; I have seen too many patients improved by the employment of the most energetic stimulants, to share the opinion of this physician on the cause of ataxic fever." Some years after this, in the third edition of his *Examination of Medical Doctrines*, Broussais says, that the foregoing declaration *was forced* from him by his respect for the opinion of Pinel, and by his fear of exposing himself to criticism! "The fact is," — he says, — "I was in error;" and instead of blushing, he glories in this refutation of himself. This "noble declaration," as Bouillaud calls it, of Broussais, would have been worthy, if not of admiration, at least of indulgence, if it had referred only to *opinions*; but how the great *reformer* managed so easily to "refute himself" on a *simple question of fact*, — of having *witnessed certain phenomena*, — and how such refutation is consistent with nobleness and honesty, is certainly not so clear to us.

necessarily resort to the use of tonics and stimulants. The object of Botal — the Sangrado of Gil Blas[22] — in his lavish and indiscriminate blood-letting, was solely to evacuate the peccant humors, which, according to his doctrine, were the causes of all disease, from the system; and thus to renew, purify, and renovate the vital fluid. In the latter part of the sixteenth century, a school of pathologists, in keeping with the mystical superstitions of the age, attributed all maladies to the influence of evil spirits — cacodemons; — chronic affections depending upon a withdrawal of the rays of the Divine Majesty, and those of an acute form depending upon an excess of the same light. Their treatment was *deduced, by a most legitimate and necessary process of à priori reasoning*, from their pathological premises, and consisted in the use of charms, amulets, and exorcisms. Broussais, before he had adopted the doctrine upon which he founded one of the chief claims to the honor of being the great medical reformer of his age — that of the local, inflammatory character of all forms of fever, — could see disease, as nature presented it to his senses; and could treat it, according to the teachings of simple experience. But not so after the adoption of his favorite dogma. He could not then see disease, as nature presented it to his senses; he could not treat it according to the results and the dictates of simple experience. Not only had be become disqualified, [212] as a careful and trustworthy observer of disease; but, worse than this, he had lost his former skill, as a safe and judicious practitioner of his art. A false philosophy of disease led him, necessarily, into an exclusive, and probably a wrong, treatment. His therapeutics was now *deduced* from his pathology; from having been *empirical*, it had now become *rational*; where he formerly saw *whatever presented itself*, he now saw only local inflammation, and for this there existed only one remedy, always local blood-letting. And so it always is — so it always has been — so it will always continue to be. The *à priori* pathologist will be an *à priori* practitioner; disqualified, just so far as the influence of his philosophy extends, both for the investigation and the management of disease.

It has sometimes happened, that the unfriendly influence of which I am speaking, has been, in a good degree, neutralized by the circumstances, that the systematist has proceeded with his *deductions* in a direction opposite to that which is usually taken; he infers the nature of disease from the effects of his remedies; he deduces his pathology, in part, at least, from his therapeutics. He first studies carefully the operation of his remedies, and on this foundation he builds up his *à priori* doctrines. His practice is really and truly *empirical*, as all practice ought to be; but in his scientific dread of this word, and of the doctrine which it designates, he hastens to render his practice, as [213] he vainly supposes, *systematic* and *rational* — to found it, as he says, upon *principles*, by connecting it with some *à priori* system of pathology. This particular form and phasis of false philosophy is strongly exhibited in the practice, and in the doctrines, of Sydenham. It is evident enough, throughout the whole of his writings, and is very expressly acknowledged in the following passages, taken from his "*Treatise of the Dropsy*."[23] The reader can hardly fail to notice his fine and true appreciation of the vicious method of procedure, which I have been endeavoring to point out. "And in reality," he says, "I am fully persuaded, that nothing tends more towards the forming a true judgment of this," —

the indication of cure, — "than an accurate observation of the natural symptoms of diseases, and the medicines and regimen which appear from practice to be beneficial or detrimental. From a careful comparison of all these things together, the nature of the distemper appears, and the curative indications are much better, and more certainly deduced, than by endeavoring to find out the nature of any determinate concrete principle of the body, to direct myself by. For the most curious disquisitions of this kind are only superficial reasonings, artfully deduced, and clothed in a beautiful dress, which, like all other things, that have their foundation in the fancy, and not in the nature of things, will be forgot in time; whereas, those axioms, which are drawn from real facts, will last as long as nature [214] itself. But though all hypotheses founded in philosophical reasonings are quite useless, since no man is possessed of intuitive knowledge so as to be able to lay down such principles as he may immediately build upon, yet when they result from facts, and those observations only which practical and natural phenomena afford, they will remain fixed and unshaken; so that though the practice of physic, in respect to the order of writing, may seem to flow from the hypotheses, yet if the hypotheses be solid and true, they in some measure owe their origin to practice. To exemplify this remark: I do not use chalybeates, and other medicines, that strengthen the blood, and forbear evacuants in hysteric disorders, because I first took it for granted, that these complaints proceed from the weakness of animal spirits; but when I learnt, from a constant observation of practical phenomena, that purgatives always increased the symptoms, and medicines of a contrary kind ordinarily quieted them, I deduced my hypothesis from this, and other observations, of the natural phenomena, so as to make the philosopher, in this case, subservient to the empiric. Whereas, to have set out with an hypothesis, would have been as absurd in me, as it would be in an architect to attempt to cover a house before he had laid the foundation, which only those who build castles in the air have a privilege of doing, as they may begin at which end they please." But it is not difficult to see, that even in [215] this modified and comparatively harmless form, the influence of which I am speaking is still, so far as it goes, unfavorable to the best and true interests of practical medicine. The hypothesis, or doctrine, in regard to the nature of disease, however cautiously and exclusively it may have been derived, from the observed effects and action of remedies, is suffered, unconsciously, perhaps, but almost unavoidably, to react upon our method of treatment, and in this way to mix itself with and to influence the practice itself. And even in the case of Sydenham himself, excellent and judicious practitioner as he was, it is quite evident, that this reaction was felt. Neither was he entirely free from the greater error, which he sees so clearly, and so strongly condemns, in others. He had his own *à priori* theory, made up of peccant matter, concoctions, commotions, and effervescences; in consequence of which, and of the other and less important fault, of which I have spoken, he was not so good a practitioner as he would otherwise have been.

It has been fortunate for the interests of humanity, so far as these interests are connected with the science and art of medicine, that the bad influences, which I have mentioned, have been much less felt by the great body of general practitioners, than by the few learned and speculative men, who have been the founders of medical sects, and by the immediate and zealous disciples. Although these influences have un-[216]questionably found their way, to some extent, into the general mind of the profession, their unfriendly effects have been comparatively limited and feeble. At any rate, it is safe to say, that they have done vastly less harm here, than amongst the authors and special partisans of the several doctrines, which have produced them; and that the number of these has generally been small, in comparison with the great mass of practical physicians. And full as the world is, and always has been, of ignorance and credulity, let us do it the justice to say, that its own observation and good sense have generally been sufficient to set it right in the matter before us; and although it rarely fails to run after each successive doctrine in medical science, hotly and blindly enough, for a time, neither does it fail, pretty quickly, in most instances, to grow weary with the chase, and to return again to the safer and beaten track of its old and better ways.

I have now a few words to say upon the evil influences of the *à priori* doctrines, or hypotheses, upon the interests and the advancement of the science itself of medicine. They have always constituted, and they still continue to constitute, the one great obstacle to this advancement. They have been the principle cause of the slow and uncertain progress of science. The sprit of the false philosophy which gives rise to them is utterly destructive of all solid and genuine progress. It is the same spirit, precisely, which kept the physical sciences so long in their infancy, and [217] which prevented their growth and development for so many centuries. So long as the spirit of this philosophy maintains its ascendency, there cannot be, in the very nature of things, any considerable degree of progress, or improvement; and this, for the very simple and manifest reason, that the powers, by which this progress and improvement are to be wrought, are all misapplied and misdirected. The goal can never be reached, for the good and sufficient reason, that the race is in the wrong direction. The attention is called away from the only legitimate objects of inquiry, and turned upon those which are in themselves wholly barren of any positive or valuable results. The senses are shut up, or obscured, or perverted; and the mind, instead of confining itself to the analysis and arrangement of appreciable phenomena and their relationships, concentrates and wastes all its energies in the construction of ingenious but idle hypotheses, which it palms first upon itself, and then upon other minds, as sound doctrines, or established principles.

So far as medical science has any just title to the appellation; and do far as medical art possesses any rules, sufficiently positive to be worth anything, it is owing, exclusively, to the diligent, unprejudiced, and conscientious study of the phenomena and relationships of disease. The sole tendency of every departure from this study, — the sole tendency of every attempt to refer these phenomena to certain unknown and assumed conditions, for [218] the purpose of rendering them *rational*, has been to hinder the progress and improvement of the science and the art. So has it ever been, so will it ever be. Here, as elsewhere, it is a straight and narrow way that leads to the truth, and however few there may be that find it, there is no other. Let no man deceive himself. The science of medicine has reached its present position, only by the labors of those who have studied the phenomena, and their relationships, of which the science consists. No man has contributed anything to its advancement, who has not *added something* to our positive knowledge of these phenomena and relationships, or aided in pointing out the only true methods of reaching this knowledge. By this inexorable test, and by no other, must every claim and every pretension be tried. In this court, it is not by his *faith*, but by his *works*, and by *these only*, that every man is to be judged. Not, what do you believe? — not, what ingenious or plausible hypothesis have you framed? — not, what supposition have you formed? — not, how do you interpret, or account for, this fact or phenomenon? — but, *what have you done? — what have you seen? — what new phenomena and relationships have you discovered? — or, what old ones have you rendered more intelligible and positive than they were before?* These are the questions which every man is to answer. And the future progress of science and art is subject to the same conditions, which have attended it thus far. People who talk, in [219] *pompous* but foggy phraseology, about what they complacently enough term the *loftier regions of philosophic thought*, and who are pleased at the same time to look down contemptuously from their imaginary elevation upon the labors of the diligent searcher after facts, will find, that these facts, few and humble as they may seem to be, and not the high speculations of the reasoner, will constitute the acceptable offering on the altar of science. Men who declaim about the importance of *principles*, and in the same breath, speak disparagingly of the dry and barren details, as they call them, of observation, will find, after all, that there are no *principles*, which have any legitimate right or claim to this character, or this appellation, excepting those, which consist, exclusively, in *these details themselves*. The *fact-finder*, and a *fact-analyzer*, is the only true contributor to the advancement and the improvement of medical science.[e] Philosophers, as they [220] are falsely called, may philosophize — speculators may speculate — sytematists may systemize — reasoners may reason — interpreters may interpret — dreamers may dream, and see visions, all to no purpose. Science consists here, as elsewhere, in appreciable phenomena and relationships, classified and arranged, and in nothing

[e] A very distinguished American author and teacher discourses in this wise upon the subject of the text. "To deny its utility," — that of theorizing, — "is to clip the wings of genius, to banish invention from the science, and to consign it over to the dull registering operations of memory alone. . . . . Can we consent to this degradation? As well might we compare the mere flutterings of the meanest and the most grovelling bird with the bold and well-sustained flight of Jove's own imperial eagle, as these slow processes of a vulgar intellect, by which facts are collected or observed, with the vigorous sallies of speculative genius, which seize truth, as it were, by intuition, and reveal it in a burst of light of celestial brightness."

What an appropriate and beautiful *pendant* to the foregoing picture, the materials of which were derived from the fertile and florid fancy of the artist, is the following truthful and sober sketch from nature, by the hand of a genuine master. "Shall we dignify," — says Sir Gilbert Blane, — "with the title of science the absurd positions of Pitcairn; the puerile and shallow hypotheses of Boerhaave and Sylvius; and deny it to those solid and applicable truths, the fruits of chaste observation and sober experience, ascertained by those methods of induction which it was the great aim of Bacon to recommend and introduce, as the only parent of legitimate, substantial and useful knowledge? *The truth seems to be, that a higher order of intellect, a more rare and happy genius, a more correct and better tutored understanding, is required to elicit practical truths by observation, than to coin theories.*" *Elements of Medical Logic.*

else; and only they contribute to its improvement, who make some additions to the extent, or the perfection, of our knowledge of these, its sole elements and materials.

Notwithstanding all this, I wish to say, as I said in relation to physical science, that I have no disposition utterly to reject and abjure all efforts to interpret the phenomena and relationships of the science of life. It is difficult, in some instances, to abstain from these efforts. I insist, only, that these interpretations, when they are indulged in, shall be regarded *as nothing but conjecture, more or* [221] *less plausible and probable*; and that they shall in no case, and under no circumstances, be admitted, or received, as essential elements of medical science. Let me illustrate these general remarks. There are many individual diseases, a part of whose natural history is constituted by a certain number, or series, more or less definitely fixed, of morbid processes, or phenomena. These processes, or phenomena, although very dissimilar amongst themselves, often succeed, or accompany each other, with great constancy and regularity. Now, it is not only very natural, but in no way incompatible with the laws and conditions of a sound medical philosophy, that we should endeavor to comprehend, in a certain qualified sense, and to interpret, the nature of the connexion between these associated series of processes and phenomena. The acute phlegmasiæ, for instance, are almost always attended by that general morbid condition, to which the term *inflammatory fever* is applied. Here are two separate and distinct processes, one of them local, and the other general, united to constitute these diseases; and the nature of the connexion and relationship between them is a legitimate subject of investigation. Is the local inflammation to be regarded as the sole, original cause and occasion of the general disturbance of the economy — the fever — or, on the other hand, is this general disturbance to be regarded as the primary disease, the local inflammation being only the result and consequence of [222] the latter? And similar inquiries may be properly made, in regard to the character of the connexions and relationships between many other pathological conditions, and between these conditions, on the one hand, and their causes and their modifiers, on the other. I am willing to go further than this, even, and to admit within the pale of legitimate inquiry, speculations of a purely hypothetical character. When these speculations are distinctly and clearly seen to be such, and are dealt with rigorously *as such*, they are at least harmless; and I do not know that they may not sometimes be of some utility, as convenient means of expressing the relations and connexions of certain phenomena. One of the finest modern instances, that I have met with, of this legitimate appreciation and estimate of its true functions and position, is to be found in Dr. Henry Holland's remarks "on the hypothesis of insect life, as a cause of disease."[24]

The long conflict between the principles of the ideal and the demonstrative philosophies, which may be said to have commenced with Plato and Aristotle, resulted, so far as the physical sciences are concerned, in the final triumph, and the permanent ascendency of the latter; and this triumph secured the steady and almost uninterrupted progress, which these sciences have ever since continued to make. The like conflict, between the same principles, so far as medical science is [223] concerned, which begun with the dogmatists and the empirics, resulted, on the contrary, in the triumphant ascendency of the former; and to this circumstance is it owing, more than to any other, that the progress of the science has been so slow and uncertain. This ascendency, in one form or another, of the principles of dogmatism, it may be confidently asserted, is at length giving way to the influences of the opposite philosophy; and the indications are too numerous and too positive, to be mistaken, of their final and utter rejection, and of the substitution, in their place, of the sole true principles of medical science — THOSE OF A PURE PHILOSPHICAL EMPIRICISM.

CHAPTER XIII.

American Medical Doctrines. Dr. Rush. Dr. Miller. Dr. John Esten Cooke. Dr. Gallup. Drs. Miner and Tully. Samuel Thompson.

AS a sort of supplement to the chapter on systems and hypotheses in medicine, I wish to take a very brief notice of some medical doctrines of American growth. These are not numerous and such a history of them, as I propose to give, will occupy only a few pages; but it may possess some degree of interest for most of my readers, while it will serve still farther to illustrate the [224] general principles of my essay, on this particular subject.

The earliest and most distinguished American writer, who can be said to have promulgated anything like a formal medical doctrine, was Dr. Rush. The leading feature of this doctrine, and his favorite dogma, was that of the *unity of disease* — a dogma not new with Dr. Rush — and of which it can only be said, that it is not merely abstract, gratuitous, and unintelligible, but in direct and manifest opposition to all common sense, to all true philosophy, and to all correct observation. Its character was very clearly exhibited, and its numerous and palpable absurdities very thoroughly exposed, in the *Preliminary Discourse* of Professor Caldwell, prefixed to his edition of Dr. Cullen.[1]

Dr. Rush had, also, a special doctrine, or theory, of fever; to which he seems to have attached much importance; and which he seems to have regarded as very sound, logical, and philosophical. It may be safely said, I think, that in the whole vast compass of medical literature, there cannot be found an equal number of pages, containing a greater amount and variety of utter nonsense and unqualified absurdity, — a more heterogeneous and ill-adjusted an assemblage, not merely of unsupported, but of unintelligible and preposterous assertions, than are embodied in his exposition of this theory. The theory is not made up of any coherent and consistent materials, and it would be [225] impossible to analyze and examine it in less space than itself occupies. Its leading ideas, however, as far as they can be got at, and separated from the confusion and obscurity in which they are involved, seem to be these; — that there is but one exciting cause of fever, which is stimulus; — that there is but one fever; — that this fever is always preceded by debility; — that it is seated in the blood-vessels, especially the arteries, and consists in an irregular or *convulsive* motion of these vessels. Amongst other assertions, scattered through Dr. Rush's exposition of his theory, are the following, — that all local inflammations are the results and symptoms, merely, of general fever; — and that all disease, which is essentially a *unit*, shows itself in one of these forms, to wit, — spasm; convulsions; heat; itching; aura dolorifica; or suffocated excitement. Dr. Rush finds no less than *nineteen* distinct points of analogy between the symptoms of fever, and convulsions in the nervous system; nearly all of them as far-fetched and shadowy, as can well be imagined. They are adduced as proofs of the truth of his doctrine, that fever consists in a *convulsive* action of the arteries. As specimens of the character and cogency of these proofs, I will cite six, not because they are more absurd than the remaining thirteen, but because they are, most of them, stated in fewer words. "Do convulsions in the nervous system impart a jerking sensation to the [226] fingers? So does the convulsion of fever in the arteries, when felt at the wrists." "Are convulsions in the nervous system attended with alternate action and remission? So is the convulsion of fever." "Are nervous convulsions most apt to occur in infancy? So are fevers." "Do convulsions go off gradually from the nervous system, as in tetanus, and chorea sancti viti? So they do from the arterial blood-vessels in certain states of fever." "Do convulsions in the nervous system, under certain circumstances, affect the functions of the brain? So do certain states of fever." "Do convulsions in the nervous system return at regular and irregular periods? So does fever." "A calm," says Dr. Rush, "may be considered as a state of debility in the atmosphere." In illustration of the impossibility of classifying diseases, he asserts, that "pulmonary consumption is sometimes transformed into a headache, rheumatism, diarrhœa, and mania, in the course of two or three months, or the same number of weeks." Such stuff as this constitutes the staple of Dr. Rush's theory; it is dignified with the glaring misnomer of "*Outlines of the Phenomena of Fever*;" and the exposition is closed with this couplet, than which, certainly, nothing could have been more appropriate:

> "We think our fathers fools; so wise we grow,
> Our wiser sons, *I hope*, will think us so."

Incredible as it may seem to be, it is nevertheless true, that Dr. Rush, in speaking of these so called [227] *phenomena*, some time subsequent to their first promulgation, says, — "he conceives the doctrine of fever that he has aimed to establish rests upon facts only, obvious not only to the reason, but, in most instances, to the senses!"

There can be no doubt, whatever, that these hypothetical fancies of Dr. Rush produced their natural and legitimate fruits; that they acted, as is always the case, very unfavorably upon his mind, and diminished, to an incalculable extent, the actual results of his scientific life. His speculative doctrines in regard to the nature of disease indisposed him to a careful and discriminating study of its phenomena and relationships, and in a great degree disqualified him for such study. They obscured his perceptions, and clouded his judgment. Worse than this, his false philosophy of disease was suffered to influence his practice, rendering this, also, more exclusive, and faulty, than it would otherwise have been. He expressly states, that the doctrine of life being a "*forced state*" was the foundation of many of his principles and modes of practice! He says, also, that his theory of life "discovers to us that the cure of all diseases depends simply upon the abstraction of stimuli from a whole or from a part of the body, when the motions excited by them are in excess; and in the increase of their number and force, when these motions are of a moderate nature."

It does not enter into my purpose to endeavor [228] to analyze, and estimate, either the intellectual character, or the scientific works, of Dr. Rush. I can only say, that if we were to judge him by his medico-doctrinal writings, it would be difficult to understand the secret of his great celebrity. They are vitiated by almost every fault of which such writings are susceptible; and they are often disfigured by incongruities, and violations of good taste. Their influence upon the general philosophy of medical science, in this country, has been extensively and altogether bad. They have helped to diffuse and strengthen all those fundamental errors, which it has been on of my chief objects in this essay to expose and remove. They have tended to call off the professional mind of the country from the pursuit of the only legitimate objects of medical science, and to lead it into the barren and foggy regions of *à priori* reasoning, and hypothetical conjecture. Dr. Rush did not accomplish a tenth part of the good, that he might have accomplished; he did not make a tenth part of the solid and valuable contributions to medical science, that he might have made, but for the disturbing and disastrous influences, of which I have been speaking, upon his own mind; and the same thing is true of his numerous disciples, who adopted the body, or imbued the spirit, of his doctrines. But notwithstanding all this, Dr. Rush was an honor and an ornament to this country, and his profession. He did good service in the cause of medical science; his vindication of the non-[229]contagiousness, and the domestic origin of yellow fever was complete and triumphant; and his history of the terrible epidemic of that disease, in 1793,[2] can never cease to be read with an instructive and solemn interest. As a false religion cannot wholly blot out and destroy all the better attributes of the soul; so, in a mind, constituted and endowed like that of Dr. Rush, a false philosophy is not able to gain so entire and ascendency, as to vitiate and enfeeble all its powers.

Another theory of fever was propounded by Dr. Edward Miller,[3] of New York, a contemporary and friend of Dr. Rush. According to this theory, all fevers are, essentially and primarily, local diseases; depending principally upon inflammation, or some other disturbance, of the stomach. This theory is very nearly identical with that of Broussais, started many years afterwards, constituting one of the chief elements of the *physiological doctrine*, as it was called, and one of the strongest claims of its author to usefulness and distinction. The honor of having *dis-essentialized* fevers, and of having converted them from obscure and indefinite morbid conditions, into local inflammations, seated mostly in the gastro-intestinal mucous surface, is not prized so highly, by any means, as it was a few years ago, when it constituted, in its wearer's estimation, and in that of his disciples, the greenest leaf in the chaplet on the brows of the then great *reformer*; but, such as it is, really belongs more to Dr. Miller, than to Broussais. In the hands of Dr. [230] Miller, as well as in those of Broussais, it would have been a harmless speculation enough, but for the usual and natural consequence of all such systems, — a pseudo-rational and *à priori* method of practice.

The most formal and elaborate American medical doctrine is that of John Esten Cooke, M. D. for many years a teacher of medicine, of some celebrity, in two of the Western schools of the United States. For the following abstract of this doctrine, I am indebted to a *Treatise on Pathology and Therapeutics*, published by Professor Cooke, in 1828. I am not aware that the doctrine has undergone any important changes since its original promulgation.

Dr. Cooke, as in duty bound, and in strict keeping with the uniform and established usage in all similar cases, commences his work with some very excellent remarks on the general prevalence, and the great dangers, of hypotheses in medicine; and on the importance of adhering closely to facts. "Long since convinced," he says, "that experiments and observations are the only true foundation of knowledge, and that hypothesis is the *ignis fatuus* by which we are led astray, the author of the following pages has endeavored, in the investigation of the changes produced in the system by the remote causes of disease, carefully to adhere to the above-mentioned," — that is, the Newtonian, — "method of philosophizing." And again, — "Accustomed [231] from the natural turn of his mind, as well as from the course of his education, to rest his belief on evidence alone, and to receive as true nothing not thus supported, he could not assent to theories built on round assertion, without the shadow of evidence to support them." The irreconcilable and flagrant inconsistency between such sentiments and the subsequent body of the *Treatise* would strike us as singular and unaccountable, but for its almost universal occurrence in the writings of medical systematists.

The pathological theory of Dr. Cooke is very easily stated; there is no obscurity about it, and it is sufficiently simple and coherent to satisfy the most rigorous *à priori* reasoner, or the loosest and laziest observer. The purest illustration of his doctrine is to be found in its connexion with fever; and it consists of the following elements. There is but one fever. Fever is always preceded by weakened action of the heart. The immediate consequence of this weakened action is a diminution in the quantity of blood sent into the arteries, giving rise to "feebleness of the pulse, paleness and coldness of the surface, diminished bulk of the external parts, shrinking of the features, and shrivelling of the skin." Another necessary consequence of the preceding condition of the heart and arteries is an accumulation of blood in the venous system, and especially in those portions of this system, which are more distensible than others and less protected by valves. These portions are [232] constituted, particularly, by the veins of the liver, and of the other abdominal viscera. The accumulation of blood in these and in other portions of the venous cavity gives rise to nearly all the symptoms and phenomena of all diseases; amongst which Dr. Cooke enumerates, in his table of contents, the following; — pulsations of the vena cava in the abdomen; pulsation in the breast, occasioning the feelings called palpitations; beating in the head, sometimes heard by the patient; shortness of breath, enlargement of the liver; debility of the muscles; serous effusions into different cavities; hemorrhages; increased and decreased secretions; dark color of the blood, countenance, and passages; high colored urine; increased and decreased secretion of the gastric fluid, and consequent variation in the power of digestion; convulsion, stupor, &c.; convulsive agitations of the body in ague, driving the blood from the external muscular parts towards the heart, and so on. All pain is also made dependent on this mechanical distention of one or more of the veins. Pain in the back, for instance, so common in the early period of fevers, is attributed to the great distention of the large venous plexus of the mesentery. Some of the foregoing effects, with other coöperating influences, result, after a time, in occasioning an increased quantity of blood to be poured into the heart, by which this organ is stimulated, the arteries refilled, and the venous plethora removed. But the increased action of the heart, thus occa-[233]sioned, is greater than it is able to sustain; its action becomes feeble again, and this feebleness is again followed by all the above-mentioned consequences.

The remote causes of all diseases produce their effect by weakening the action of the heart. The two most important of these are carbonic acid gas, and cold. The former is the remote cause of all summer and autumnal fevers; the two, combined, constitute the remote cause of all winter epidemics.

Such are the outlines of the theory of Dr. Cooke, in its application, particularly, to the subject of fever. There is no essential variation, however, of the theory, in its application to all disease. I have endeavored to state the doctrine clearly and fairly; but lest some of my readers should feel a very natural skepticism as to the entire truthfulness of this statement, I will quote, literally, a paragraph from the second volume of Dr. Cooke's *Treatise*. It is in the following words: — "The effects of accumulation of blood in the venous cavity, variously combined, and sometimes one and sometimes another more prominent than the rest, constitute most of the diseases to which man is liable." Dr. Cooke makes all the phenomena of inflammation dependent, exclusively and immediately, upon debility of the arteries and veins where it is seated; so that they are distended by the blood driven into and accumulated in them by the heart. [234]

I hardly need say, that I have no intention of entering into any analysis, or examination, or this most extraordinary doctrine. Everybody, not already struck with the hopeless blindness resulting from these pseudo-philosophical phantasms and illusions, must see, at once, how utterly gratuitous all the assumptions are, upon which it rests, and in which it consists. Is there any evidence, whatever, of the uniform *weakened action of the heart*, preceding all diseases, which lies at its foundation? Not the slightest. Is there any evidence, that this assumed condition of the heart, even if it did exist, must produce the effects so confidently attributed to it? Not the slightest. Is there any evidence, except in a small number of cases, of an undue, mechanical accumulation of blood in the venous system? Not the slightest. Is there any evidence, even if this assumed accumulation were actually present, that the effects so unhesitatingly assigned to it, are its legitimate or necessary effects? Not the slightest. Is there any evidence, that the remote causes of disease always, or generally, act by weakening the energy of the heart? Not the slightest. Is there any evidence of the existence, and the assumed *modus operandi* of carbonic acid gas, as the one chief cause of epidemic fever? Not the slightest. And so on, from beginning to end. Not only do all the propositions of the theory consist of what the author calls *round assertions*, unsupported by the veriest shadow of proof; but most of them are either wholly incom-[235]prehensible, or in direct opposition to all observation and pure experience. The entire system is one, of the purest *à priori* speculation, and that, too, of the absurdest and most improbable character. Its pathology is unmixed mechanism, — the body is only a frame-work for a hydraulic apparatus, and its diseases consist, almost exclusively, in derangements in the mechanical working of this apparatus; the existence of strictly vital properties being hardly recognized even, anywhere in the system.[4]

Dr. Cooke's *therapeutics* is of course *deduced* from his *pathology*. It is founded on the following four indications: 1. To remove the remote causes, which may be still operating on the heart, weakening its action; 2. To excite and support the action of the heart; 3. To reduce the quantity of blood in the venous cavity; 4. To reduce the action of the heart, produced by the press of blood from the venous cavity, if it exist. It is not my purpose to give a detailed account of the methods by which Dr. Cooke fulfils these indications. The leading and prominent peculiarity of this therapeutics corresponds to one of the principal elements of his pathology; and consists in removing the sanguineous engorgement of the liver, by means of active cathartics, especially calomel, aloes, and rhubarb. The test of their success, in accomplishing this object, is to be found in the qualities of the discharges; and it is known to be accomplished, when these are consistent, and when their color is yellow, green, dark, or black. Bleed-[236]ing is another means sometimes employed for the same purpose; but it is always used simply as a mechanical operation, for emptying the venous cavity of a part of its superabundant fluid.

The ridiculousness of some portions of the therapeutics of this system might well excite our risible emotions, if these were not swallowed up by reflections of a more grave and serious character. The follies of German uroscopy[5] are outdone by those of practical *vena-cavaism.* Its votaries seem to have forgotten, that there is any organ in the body except the liver, and, in the management of disease, the only important points to be determined are, what is the *color*, the *consistence*, the *odor*, and the *quantity* of the stools. Diagnosis is wholly discarded, as a matter merely of idle curiosity, and of no practical importance; and prognosis is founded, almost exclusively, upon the character of the alvine evacuations. If these are *bilious*, as it is termed — if they are consistent, and dark-colored — everything is going on well, and the prognosis is favorable. The Cookeite would be utterly at a loss, in regard to the state of his patient, if he should be deprived of the aids which are furnished him, by a daily and nightly inspection — ocular and nasal — of the stools. They constitute his guiding star, his rudder, and his compass; they shed a clear light on all his pathway, which, but for them, would be darkness and uncertainty itself . The language of his sect, as usually happens in similar cases, has [237] passed into the popular tongue; and we hear from all invalids daily and hourly complaints, that the liver is *locked up* — that the liver is *torpid* — that the liver does not "*act*" — and so on. Almost every ailment to which the body is subject — functional or organic — trifling or grave — chronic or acute — is immediately referred to this ubiquitous and autocratic organ; all and each of these ailments can be removed in only one way, — by inducing the liver "*to act*;" and this can be accomplished with certainty, only by one infallible remedy — calomel. This substance is proclaimed to be, not only the most efficacious and important article of the materia medica, but, also, one of the safest and most inoffensive. It is constantly administered — on all occasions — in all diseases — and in all their stages. It has, literally, in some instances, been made an article of *daily food* — sprinkled upon buttered bread, and mixed with it before baking! I suppose it is no exaggeration to say, that there is more calomel consumed in the valley of the Mississippi and its tributaries, than in all the world beside.[a] I have [238] heard more than one extensive and unprejudiced observer express the opinion — so enormous and indiscriminate is the use of this substance, throughout this region — that, compared with the present practice, its entire expulsion from the materia [239] medica would constitute a blessing of incalculable value.

Professor Cooke's doctrine more strictly fulfils all the conditions that ought to attach to this class of intellectual products, than any other with which I am acquainted. It is quite perfect in the two leading and fundamental qualities of these systems,— namely,— comprehensiveness, and simplicity. None of its predecessors, from methodism to homœopathy, can rival it in these respects. It certainly deserves the distinction of being one of the *type-species* of the extensive family to which it belongs; it possesses, in an eminent degree, and in their highest purity and perfection, all the distinguishing properties of its tribe. The great rules, which should preside over the construction of an epic, or a drama, were never more scrupulously regarded, than have been the true principles of a complete, coherent,

---

[a] Amongst the single cases, reported in Dr. Cooke's *Treatise*, illustrating his treatment, is the following; of which, lest some of my readers should suspect me of exaggeration, in what I have just said, I will give a brief abstract in this note. If they still doubt the authenticity of the case itself, I can only refer them to the work from which I have taken it. They will find it recorded *in company with several others of a similar character!* in the second volume, between pages 242, and 254. A gentleman came into Dr. Cooke's hands, in March 1824. He had been an invalid since 1807, when he had intermittent fever, and so far as the history of his symptoms enables us to determine, was left with enlarged liver and spleen, serious gastric disease, — perhaps structural ,— and probably organic disease of the heart. He had already been treated somewhat variously and heroically, during a good part of this long interval. From March to April, he took *large quantities* — the precise amount is not stated — of calomel, rhubarb, aloes, and jalap — sufficient to procure three or four stools daily. He then removed to Dr. Cooke's residence, and took, in the course of ten weeks, 240 grains of calomel, with rhubarb and aloes. By July, he had become much worse than before, exceedingly feeble, very thirsty, and suffering with an intolerable internal burning. Between July 27th, and August 10th, — fourteen days — he took 410 grains of calomel, 270 grains of rhubarb, and 20 grains each of jalap and scammony. From the latter period, up to the end of September, he took 836 grains of calomel, 983 grains of scammony, 840 grains of rhubarb, 630 grains of jalap, and 560 grains of aloes; besides *occasional* other doses of some of the same articles. At the close of this most pregnant history, the writer of it says of its subject: "He died about the 1st of December;" and immediately adds these words: "*The case is full of instruction*." At one time Dr. Cooke admits, that he began to be perplexed; and the idea even occurred to him, that the poor martyr's health "*ought not to be risked on a theory*." He says, however, that the patient's "relapse and death, were clearly caused by improper diet, and the use of brandy."* The case is, indeed, so full of instruction, that this brief abstract of it will be quite sufficient, I am sure, to answer all my purposes, without any comments; and so I leave it.

* Ce n'est pas qu'avec tout cela votre fille ne puisse mourir; mais au moins vous aurez fait quelque chose, et vous aurez la consolation qu'elle sera morte danz les formes. — *Molière*

and consistent *medical doctrine*, in the present instance. None of the unities have been violated. It is as intelligible to the smallest as to the largest capacity; it may even be said, I think, that it is capable of being seen through, comprehended, and understood, more readily by one who is ignorant of the subtle and complicated properties and relations of the vital organization, than by one who has carefully and profoundly studied these subjects. There are no facts so stubborn as not to bend easily to its requisitions; and it embraces, with perfect facility, phenomena of the most [240] opposite qualities and character. By a most felicitous contrivance, it reduces the manifold and obscure causes of disease to a single and obvious influence; removes the supposed necessity of distinguishing between apparently different morbid affections; and then, as its crowning glory, furnishes us with means of removing these diseases, as simple as the nature of the diseases themselves. The old and common notion, that in order to cure a disease, it is first necessary to know where and what the disease is, is shown to be wholly erroneous; and the irksome and oftentimes difficult work of diagnosis is rendered entirely unnecessary. It can hardly be considered singular, that a pathological and therapeutical "*ready reckoner*" of such facile application, should have come into pretty general use; and that it should have superseded, somewhat extensively, the more complicated and laborious processes, which have been generally thought necessary, in order to arrive at safe or positive results.

A few words respecting three other medical doctrines, advocated by American writers, will complete these hasty sketches. The doctrines to which I refer are those of Dr. Gallup, Drs. Miner and Tully,[6] and Samuel Thompson. I do not know that there is anything in either of them particularly new, and they are chiefly interesting to us as striking illustrations of the general principles of the preceding chapter. It is curious to see, in the case of the two first mentioned doc-[241]trines, in what opposite directions, and how far, both from each other and the truth, the minds of men may be carried, when they break away from the moorings of a sound and positive philosophy. The three physicians, first named, were cotemporaries; they were men of active and acute minds; they were medical authors and teachers, long and industriously engaged in the investigation of disease, and they occupied essentially the same field of observation. But, notwithstanding these circumstances, — although they followed their profession, and found their patients, on the borders of the same river, suffering with precisely the same diseases, — their systems of pathology and of practice were, in almost every particular, in diametrical and unqualified opposition to each other. Leaving out of view any minor differences between them, it is sufficient to state, that they disagreed, totally, in regard to the very nature of the morbid action, in nearly all cases of disease. According to the system of Dr. Gallup, the diseases of New England, during the present century, have been, almost universally, sthenic, or inflammatory, in their character; according to that of Drs. Miner and Tully, they have been asthenic, or non-inflammatory. Precisely the same obvious morbid phenomena are interpreted by the two schools on directly opposite principles; and, as a legitimate and necessary consequence, these opposite interpretations led to corresponding opposite methods of treatment. Dr. Gallup says, — in [242] evident allusion to the practice of the rival school, — "some servile imitators of the *incendiary treatment* have been very vociferous in vindication of principles, which are capable of destroying more than the pestilence itself."[b] Bleeding is the sheet anchor of Dr. Gallup, and when he administers stimulants, in the lowest and most malignant forms of disease, — in spotted fever, scarlatina, and typhoid pneumonia, — he does so, principally, in order to prepare the system for this heroic remedy. "The lancet," says Dr. Tully, "is a weapon which annually slays more than the sword;"[c] and, again, on the next page, — "The King of Great Britain, without doubt, loses every year more subjects by these means," — depleting remedies,— "than the battle and campaign of Waterloo cost him, with all their glories." "It is probable," says Dr. Gallup, "that forty years past, opium and its preparations have done seven times the injury they have rendered benefit, on the great scale of the world."[d] "Calomel and opium," says Dr. Tully, "in acute febrile diseases, are of greater service, than all the other articles of the materia medica. There is no good physician, in full practice, who does not employ them daily."[e] "Such practices," proclaims Dr. Miner, referring to bleeding, and other depleting measures, "have been [243] the scourge and devastation of the human race for more than two thousand years;"[f] and in allusion to the opposite class of practitioners, Dr. Gallup responds, — "We are not content to speak through pages which may never reach the public eye, but with for a lengthened trumpet, that might tingle the ears of empirics and charlatans, in every avenue of their retreat."[g] But let us pause; certainly, this humiliating exhibition has been continued quite long enough; but if it helps, in any degree, to render intelligible and striking the unfriendly and disastrous influences of these hypothetical systems, both upon the general interest of science and humanity, and upon the minds and tempers of their authors, I shall be justified in having made it.

Samuel Thompson,[7] whose name I have mentioned, was the founder of what is sometimes designated the *botanical*, and at other, the *steam* system of practice. Mr. Thompson was an illiterate man, and never received even the rudiments of a medical education; his disciples are, almost without exception, men of like character; so that this American medical doctrine has never been recognized, as a legitimate member of the family to which it claims to belong; it has never received its diploma; it has never been allowed to take its degree.[h] But notwithstanding that [244] the system is

---

[b] Gallup on the Institutes of Medicine. Vol. ii. p. 370.

[c] Essays on Fever, etc. By Thos. Miner, and Wm. Tully, p. 460.

[d] Gallup on the Institutes of Medicine. Vol. ii. p. 187.

[e] Essays, etc. p. 274.

[f] Essays, etc. p. 80.

[g] Gallup on the Institutes of Medicine. Vol. ii. p. 298.

[h] A very curious feature, in the history of many of these medical doctrines, consists in their moral and social relations; each one prevailing most extensively in certain pretty well marked classes of the community. Thompsonism, for instance, like certain forms of religion — Mormonism, and Millerism — finds the greatest number of its adherents amongst the least educated portions of the people; while homœopathy, on the other hand, is received with special unction and favor, by the more intelligent and better educated classes; and particularly by persons, the tendencies of whose minds are towards ultra and abstract principles in politics and morals, and rational mysticism in religion. A non-resistant,

thus repudiated by the regular member of the profession, and its practitioners denounced, as charlatans and pretenders, a very brief statement of its peculiarities will show, that its general philosophy is, in all respects, as sound as that of those, which have already occupied our attention. It possesses most of the elements of these doctrines; is constructed on similar principles; rests upon the same foundation; is as strictly and rigorously *inductive*, as thoroughly *Baconian*, as they are. Considered as a whole, it partakes, to be sure, somewhat of a mongrel character; its pathology and physiology having the classic physiognomy of the old Greek philosophy, while its therapeutics is a compound of Indian and Yankee empiricism. In its general construction, although it may be inferior in artistic and elaborate simplicity, and in the congruity of its several parts, to the doctrine of Dr. Cooke, it is certainly superior, in these respects, to those of Dr. Rush, and Hahnemann; while it is altogether more reasonable, and more intelligible, than either [245] of these latter. Mr. Thompson found, he informs us, "after maturely considering the subject," that the human body is composed of the four elements, earth, water, air, and fire; the solid parts being composed of the two former, and the fluids, of the two latter. Heat, he found, was life; and cold, death; all disease consisting, essentially, whatever might be its form, in a diminution of heat, and depending upon obstructed perspiration, as its exciting cause. Such are the *general principles*, as Mr. Thompson calls them, of his physiological and pathological doctrine; from which he *deduces* his appropriate and *à priori* method of treatment; thus freeing it from the reproach of being *empirical*, merely, and rendering it sufficiently *scientific* to satisfy the most thorough-going *rationalist* therapeutist. Mr. Thompson's principle remedies were lobelia, the steam bath, and cayenne pepper; all used for the purpose of fulfilling certain *rational indications*; the first, *to remove obstructions*, and the two others, *to keep up the internal heat*, and so to counteract the tendency to cold, which is death. Besides this fundamental and *philosophical* affinity between the system of Thompson, and those of other medical *doctrinaires*, there are, also, many other points of analogy between them. Mr. Thompson, like Dr. Rush, insists upon the essential *unity of disease* — a very useful and convenient *principle* — since it enables him, like the system of Dr. Cooke, to get rid of the embarrassments and difficulties, which [246] so frequently attend the positive diagnosis of disease. With Tully, Miner, and Hahnemann, he denounces bleeding, which, he says, "always reduces the heat, and gives power to the cold." Homœopathy coolly refers nearly all disease to the *itch*; Thompsonism, just as coolly refers them to *canker*; and it would not be an easy matter to determine which of these two *pathological principles* is most in accordance with scientific observation, and sound common sense! Mr. Thompson's deportment, in his writings, towards the *regular doctors*, as he calls them, is characterized by precisely the same kind of courtesy and fairness, which mark that of most of the rational systematists towards each other. He has written a *Treatise on the Laws of Life and Motion*,[8] made up of sundry physiological and

---

transcendentalist, and Grahamite, makes the most devoted disciple, and the stanchest advocate of homœopathy.

pathological commentaries, constituting what may be called the *Institutes* of his system; which, if less learned and voluminous than those of some of his countrymen and *confreres*, are altogether more intelligible, both in their matter, and in its exposition. Thompsonism, finally, vindicates its claims to our regard and acceptance, by an array of successful cases, and of marvellous cures, quite as authentic and imposing, as those of *metallic-tractorism*,[9] or homœopathy. I may be allowed to hope, that my countrymen will properly appreciate, both the sense of duty, and the professional patriotism, which have prompted me thus to vindicate the rights of a medical theory, which has been rather [247] cavalierly treated, to the honor and dignity of an equal position amongst its somewhat arrogant and supercilious kindred.

But the spirit of a false philosophy, in this country, has manifested and embodied itself, less in the formal construction of entire and consistent systems, or doctrines, like those of the methodists, the contro-stimulists, of Brown, of Hahnemann, of Cooke, than in modes and forms of a different character. This spirit pervades almost the entire science of medicine amongst us, as it does amongst the British, from whom we have principally derived it, and shows itself in many and miscellaneous forms, by a general departure from the true principles of medical philosophy.

One of the most common of its manifestations is to be found in the general misapprehension, which exists, as to the very nature and constituents of what is called a *principle*. A great deal is said, by our own writers and teachers, about the importance and necessity of what they call *general principles* in medicine. They declaim about the arid barrenness[10] of experience without *fixed principles*; and warn us of the dangers and absurdities of *empiricism*. They say, that medicine can become a *science*, only through the aid, and by the agency, of these *principles*. But, notwithstanding all this, it is quite evident, that they look upon a *principle* as something else, than the rigorous generalization of certain phenomena, or relationships; — something other, than the simple expres-[248]sion of a general fact. What the essential elements and conditions of one of these so called *principles* are, however, they do not tell us; and it is impossible to get at their own views with any degree of definiteness, or certainty. The idea, generally attached to the word, seems to be of a mixed quality; according to which, a principle in medicine is the product, or result, of a large amount of *à priori* reasoning, — a great deal of what is called *induction*, — employed and expended upon a small number of facts, or phenomena; and then erected into an arbitrary standard, by which all other facts and phenomena are to be measure and tried, — an absolute law, by which they are to be judged and governed. If this is not the meaning attached to the word, — if these are not the attributes assigned to the thing itself, — I am wholly unable to comprehend what that meaning, and these attributes, are.

Another form, in which this spirit shows itself, is that of the almost universal mania, which exists, for explanations and interpretations of all phenomena and their relationships. The dominant feeling, in the American medical mind, seems to be, — not what *are* the facts and their relationships? — to what extent, and with what degree of positiveness and accuracy, have they been ascertained? — but *why* are these facts and relationships such as is alleged? And *how* are they so? A vastly greater degree of importance is often attached to the possible, though perhaps wholly [249] unattainable *why*, and *how*, and *wherefore* of the phenomena, than to the phenomena themselves; and in strict conformity to the requisitions of this strange philosophy, in many cases, unless some plausible, or satisfactory answer can be given to these questions, the very existence of the phenomena themselves is coolly and complacently denied! We have practically reversed one of the sayings of John Hunter, — "*Don't think, but try*;" and adopted its opposite, — *Don't try, but think.*

CHAPTER XIV.

PROPOSITION FIFTH

DISEASES, LIKE ALL OTHER OBJECTS OF NATURAL HISTORY, ARE SUSCEPTIBLE OF CLASSIFICATION AND ARRANGEMENT; THIS CLASSIFICATION AND ARRANGEMENT WILL BE NATURAL AND PERFECT JUST IN PROPORTION TO THE NUMBER, THE IMPORTANCE, AND THE DEGREE OF THE SIMILARITIES AND THE DISSIMILARITIES BETWEEN THE DISEASES THEMSELVES.

The principles and conditions of nosological arrangements. These arrangements necessary. Classifications in botany. The artificial and natural methods. Are diseases legitimate objects of classification? Defects of nosological systems. Examples of natural groups, or families. Exanthemata. Fevers. Phlegmasiæ. Cancer and tubercle. Neuroses. Definitions.

I HAVE already remarked, in the short chapter on the subject of classification and arrangement in [250] the physical sciences, that as I was not writing a treatise on these sciences, my only object was to state, simply, and as briefly as possible, the principles on which this classification and arrangement should rest. I wish to make a similar remark, at the outset of the present chapter; — I am not writing a treatise upon natural history, in general, nor upon that portion of it, which is concerned, especially, with diseases; and it does not fall within the scope of my essay, to go into a full consideration of the arrangement and classification of the objects and phenomena of natural science. My single object is, to endeavor to ascertain, and to point out, as clearly as the present state of science will enable me to do so, the principles and rules, which ought to govern us, in the arrangement and classification of those phenomena and relationships, exclusively, which constitute what are called diseases.

Some arrangement of this sort is a matter of as much convenience and necessity, in this branch of natural history, as in any other. Individual diseases, or forms of disease, are very numerous; many of them have certain resemblances to each other; and the suitable disposition of these diseases, in classes, or groups, or families, is just as essential to the character and completeness of medical science, as a like disposition of plants and of animals is, to the perfection of the sciences of botany and zoology. This necessity has always been felt; and it has given rise, from time to time, [251] to the various classifications of diseases, which have constituted the systematic *nosologies*.

Perhaps we shall be better prepared to see, clearly, the rules, which ought to govern us, in the construction of a system of nosology, if we first look to the principles, which have been adopted, in the arrangement of the objects of some other branch of natural history; and amongst these, there is no one so well suited to our purpose, as botany. There are more points of resemblance, so far as this particular matter of classification is concerned, between the science of botany, and the science of pathology, than there are between the latter and any other department of the natural, or classificatory, sciences. Without entering at all into the history of botany, it is sufficient to say, that in the classification and arrangement of plants, two principal, or leading, systems have been adopted. One of these is the *Linnæan*, or what has been called the *artificial*, system. The division of plants into their primary groups, or classes, according to this method, is founded, with a single exception, upon the number, situation, relative length, and so on, of certain parts of the flowers, called stamens; and upon these circumstances alone. Every other element in the form of plants, every other organ, or part, is disregarded; and no notice, whatever, is taken of their internal structure, their functions, habits, or qualities. Provided, only, that plants resemble each other, in a single peculiarity of external form, [252] connected with the above mentioned parts of their flowers, they are placed together, — however dissimilar they may be in all other respects; — and however closely they may resemble each other, in their general form, in their internal structure, in their functions, in their properties, or in any other respect, they are placed in separate, and, it may be, widely sundered classes, if they are not closely allied in the arrangement of these floral appendages. And the first subdivision, into orders, still depends upon the differences and peculiarities in other parts of the inflorescence, and fructification, — as the number of styles; the covered, or naked, condition of the seeds; the relative length of the pods, and so on.

The second of these systems, first fully developed by the younger Jussieu, and called the *natural method*, founds its division of plants into classes, orders, genera, and so on, not upon any single circumstance in their external form, or character, but upon a wide and comprehensive comparison of *all* the phenomena and relationships, which unite to constitute a plant. Thus, this system begins with the study of what may justly be regarded as the most important element in the constitution of plants, — their intimate, anatomical structure, — so far as this is ascertainable. A careful and minute examination of this structure, by the microscope, and by other means, has demonstrated the existence, in some plants, and the absence, in others, of certain important organs, [253] which are called *spiral vessels*. On this anatomical difference, is founded the great, primary division of the members of the vegetable kingdom, into two classes, *vasculares*, in which the spiral vessels are present; and *cellulares*, in which they are absent. That this division is primary and fundamental, and that the element, on which it rests, is of the same character, is clearly shown by the fact, that each of these original classes is characterized by other constant and uniform peculiarities; amongst which may be mentioned, particularly, this — that vascular plants have distinct flowers with stamens and pistils, while cellular plants are destitute of these organs. On farther inquiry, it is found, that plants having spiral vessels, and bearing flowers, are propagated by seeds, and are hence called *phenogamous*: while the opposite class, which are destitute of spiral vessels, and of flowers, are not propagated by seeds, but by bodies called *sporules*, and are hence called *agamous*. It is impossible not to see, at once, that a division, marked by such important peculiarities in the internal structure, the external appendages, and the mode of propagation, must be *primary* and *essential*.[1] On pushing this examination still further, it is found, that the class of vasculares consists of two sub-classes, as distinct from each other, as the vascular class is from the cellular; the difference between them depending upon anatomical structure, and mode of growth. The plants, belonging to one sub-[254]division, grow by the addition of successive layers to the outside of their trunks and branches; and for the protection of the newly grown and delicate layer, thus annually formed, they are furnished with a coat, or envelope, called the bark. These concentric tubes of woody substance are firmly held together, by a tissue passing through them, or connected with them at right angles, called medullary rays; and the common central axis of the tubes is occupied by a cellular substance called pith. This subdivision of vascular plants is called from the mode of growth, which distinguishes it, *exogenous*. Plants, belonging to the other subdivision, grow by additions made to the inside of the trunk, or stem; they have no bark, nor does their internal structure consist of concentric layers, united by medullary rays. It is thus seen that amongst the characters, upon which is founded the primary and most important division of plants, into classes, is that of their *anatomical structure*, an element which the system of Linnæus wholly overlooks. These fundamental differences, in the internal organization of plants, manifest their importance in the vegetable economy, by various other modifications, which accompany them in the several classes. The two great classes of *exogens* and *endogens*, for instance, may be distinguished from each other, simply by an inspection of the veins of their leaves; so that from this apparently trifling circumstance, — the venation of the leaf, — the anatomical structure, [255] the mode of growth, and the number of cotyledons attached to the seed of the plant to which the leaf belongs, may all be determined. Besides the primary divisions, of which I have already spoken, there are still others, of a very general character, founded upon some single and prominent peculiarity, before we arrive at the final orders, — the *little family groups* themselves. For instance, the great sub-class of *exogenæ*, is divided into two tribes, distinguished, through all their wanderings and ramifications, from each other, by the single circumstance, that the seeds, in one tribe, are enclosed in what is called a *pericarpium*; while, in the other, the seeds are destitute of this envelope. The same sub-class is again divided into smaller tribes, distinguished from each other, not by anything in the dispositions of the seeds, but by certain peculiarities in the form, and arrangement of the parts constituting the flowers.

The subsequent divisions and subdivisions depend upon a great number and variety of peculiarities; and these peculiarities are derived from an examination of *all* the properties, qualities, and relations of the several plants, and groups of plants, which are united to form them. Lindley, in his *Introduction to the Natural System of Botany*, enumerates more than twenty distinct parts, or organs, which are made use of, in determining the affinities of plants, and in establishing the natural groups, or orders, in which, according to these affinities, they are arranged. Each group, [256] or order, or family, is marked by the possession of certain characteristics, more or less numerous and striking, and consisting in peculiarities in some one, or more, of the several parts, or organs constituting the plants. A full illustration of this subject would require an amount of room larger than I can afford to it; and, in order to be intelligible, a somewhat extensive acquaintance with a great number of terms, which are strictly technical, in their use and meaning. I think I have said enough, however, to exhibit the general character of the two leading methods of classification, which have been adopted, in the science of botany; the principles upon which each is founded; and the fundamental differences between them. One of them, — the *Linnæan*, — establishes all its primary divisions upon the character of a *single apparatus*, — that of fructification, — overlooking *all the other elements*, which go to make up the natural history of plants. Its advocates claim for it the merit of simplicity, intelligibility, and facility of application in determining the position, in it own ranks, of the several individual species of the vegetable kingdom; and this merit can hardly be denied to it. But it is obtained at the expense of violating the most important rules, and the most essential conditions, of all natural and philosophical classification. It disregards the most important properties of plants, and takes no cognizance, whatever, of their most numerous and striking differences and affinities. [257] It sacrifices naturalness and comprehensiveness to an artificial and superficial simplicity. The other method, or system, — that of Jussieu, — finds its principles in a careful and thorough study of the *entire economy* of the vegetable kingdom. Its primary divisions rest upon what may be called the fundamental anatomy and physiology of plants; and after having exhausted the simpler, more uniform, and more essential elements of vegetable organization, in the establishment of these primary division ; in the constitution of its ultimate groups or families it takes into consideration, and endeavors to ascertain, the value of *all* the affinities and relations, which bind these groups together, and of *all* the circumstances, by which they are separated, and in which they differ from each other.

In turning now to the more immediate subject of this chapter, — the methodical classification and arrangement of diseases, — the first question, to be determined, is this, — *whether diseases are endowed with such properties, and so constituted, as to render them legitimate objects of systematic classification*? Are the elements, of which they are composed, sufficiently fixed and determinate, to give them an individual character; and to render them, as species, comparable one with another? — or are they mere forms, or modes of manifestation, of certain conditions, so mixed up, confused and running into each other, as to take from them all individuality? This latter is the view, which has [258] been taken by some very distinguished and very able pathologists. It occupies a prominent position in the so called *physiological doctrine* of Broussais, and his disciples. Broussais is constantly declaiming against the error and absurdity of elevating morbid conditions into distinct entities. He pours out his vehement and stormy indignation upon this *miserable ontologism*, as he is pleased to term it, without stint or measure; and he denies, without qualification, that there is any resemblance between diseases, and other objects of natural history, as subjects of systematic arrangement and classification. My own opinion, in relation to this matter, is implied in many of the doctrines of this essay, already stated in the preceding pages. This opinion is that, which is generally received, and acted upon; and it seems to me hardly worth while to enter into any formal vindication of its correctness.[2] There may be, it is true, many morbid conditions, so indeterminate in their character, or so imperfectly known to us, as not to be amenable to this process of individuation. But, certainly, this is not the case, in the great majority of instances. The most numerous and the most important morbid conditions are sufficiently marked, and distinguished from each other, to constitute them comparable objects, and to render them susceptible of being dealt with as such. The morbid conditions, generally designated by the names *apoplexy, pneumonia, pericarditis, phthisis, typhoid fever, measles, scarlatina,* and *small pox*, are just as [259] clearly and unequivocally morbid entities, or individuals, or comparable objects, as any others, whatever, in natural history; and they are just as susceptible of definition, and arrangement, as those of zoology and botany. There is no essential difference, in this respect, between the objects of pathology, and those of other branches of natural science; although the great complexity and variableness in the phenomena and relationships of the former, and the imperfection and incompleteness of our knowledge of many of these phenomena and relationships, may render the process of individualizing, arranging, and classifying them, more difficult, than this process is in other departments of natural history. [a]

It is a fact, so notorious as to stand in need, neither of evidence, nor illustration, that all the efforts, which have hitherto been made, at a formal and systematic arrangement of disease, have proved unsatisfactory, and unsuccessful; and nothing can aid us more in getting at the true principles, and the necessary conditions, of any [260] natural and successful arrangement, than an examination of the causes of this universal failure; and this can be done without going into any detailed or irksome history of the several classifications themselves.[3]

---

[a] Dr. Rush, in conformity to his hypothetical notions about the *unity of disease*, outdid even Broussais in his opposition to nosological classifications. He goes so far as to say, that, "they *degrade the human understanding*, by substituting simple perceptions to its more dignified operations in judgment and reasoning." This sentence, let me remark, by way of parenthesis, so imposing and oracular, is a very authentic and summary expression of the still prevalent spirit of medical philosophy. The reply of Professor Caldwell to the doctrines of Dr. Rush, in his *Preliminary Discourse*, already referred to, is unanswerable and conclusive; and his general remarks upon the entire subject are sound, lucid, and philosophical.

We shall find, from such an examination, that there is one original and parent cause of this want of success; — that there is one radical defect in the principle of all these arrangements, lying at their foundations, and running through and vitiating all their details. This fundamental error consists in the circumstance, that they are founded, not upon an examination and comparison of *all the phenomena and relationships*, which constitute diseases, but upon an examination and comparison of *certain particular and limited portions* of these phenomena and relationships. In these systems, diseases have been arranged to a great extent, at least, according to their similarities, and dissimilarities, *in a few particulars*; and these, perhaps, of minor importance; and not according to their general, or their most essential, affinities. The three methodological nosologies, which have been more popular and successful, than any others, are those of Sauvages,[4] Cullen,[5] and John Mason Good;[6] and although they differ, in some respects, from each other, they are all marked by the fatal defect of which I am speaking. They take into consideration only a certain limited number of the elements of diseases; these elements are arbitrarily chosen; and in many if not in most cases, they [261] are less essential, less constant, and of course, less characteristic, and distinctive, than some others, which are overlooked. Generally, these elements consist of the more prominent and striking *symptoms* of the several diseases, which are the subjects of classification; and the classification itself rests upon, and consists in, the similarity and dissimilarity of these symptoms. The systems are, indeed, what many of them profess to be, merely *symptomatic* nosologies. The seat, or locality, of diseases, it is true, constitutes a portion of their natural history too striking and important to be disregarded; and we accordingly find, that wherever this seat was ascertainable, it is often made use of, as a means of determining their position. But, almost without exception, in all these systems, some of the most essential elements in the natural history, even of these local diseases, are left wholly out of view, both in fixing their position, and in defining their characters. The most remarkable and uniform omission, of this sort, is that of the morbid processes and alterations, which *really constitute the diseases themselves*. It is hardly necessary to say, that this omission renders the specific definition of these diseases fatally defective, and utterly inadequate; or that their true position, in any methodical and natural arrangement, can never be determined under such conditions. This position can be fixed only by a comparison of *all* the ascertainable characters of these diseases with each other, and with those of other [262] diseases. Dr. Good's definition of pleurisy is in these words; — "*Acute pain in the chest; increased during inspiration; difficulty of lying on one side; hard pulse; short distressing cough.*" This is no definition of pleurisy, for the simple and obvious reason, if for no other, that the only constant and essential characters of the disease are wholly omitted. The disease may exist without any of these phenomena; and the definition is as applicable to many cases of pericarditis, as it is to pleurisy itself.

There is no occasion for pursuing this examination any further. It would advance my purpose but little, to analyze these several systems, in detail, in order to point out their minor faults — their imperfections, inconsistencies, and shortcomings. I will merely mention two other causes of these imperfections. One of them is to be found in the vain effort, so constantly visible, in all these systems, to render the arrangement of diseases *fixed* and *absolute*, like that of the phenomena and objects of physical science. It is this effort, which has led nosologists into the radical error, already so strongly insisted upon, of adopting some single, exclusive principle, or standard, of classification; in consequence of which, the classification itself became, necessarily, artificial and arbitrary. Naturalness and truth were sacrificed, to a false and fallacious simplicity. The other cause, to which I allude, is to be found in [263] the great imperfection of our knowledge of many individual diseases, at the times when these systems were constructed. *A thorough knowledge of diseases* must *precede* their classification; and even if the methodological nosologists, from Plater[7] to Good, had followed the true principles of classification, their systems must have still been exceedingly defective, from the want of this knowledge.

I do not intend, in this essay, to attempt the formation of a complete nosological arrangement. My object is, merely to ascertain, if this is possible, the right principles, and the necessary conditions, of such an arrangement; and to indicate, in a general way, the method, which must be followed for its accomplishment; and in accordance with the views of the present chapter, let us now endeavor to see what this method is, and how far it promised to be successful. Let us call some of the morbid conditions, to which the human body is subject, into our presence, and see how far, and in what manner, their differences and their affinities, their attractions and their repulsions, will lead to their arrangement, in separate classes. The result, I think, of such an examination, will be this — that nearly all of these diseases will be found to dispose themselves, at different distances, in what may be called *natural*, or *family, groups*, round certain common centres — each centre, or the circle nearest to it, being occupied by one, or more, of the *type-species* of the family to which it [264] belongs.[b] As the affinities between these *type-species*, and other diseases, diminish in number and importance, the latter will recede, farther and farther, from the neighborhood of the former; until they finally fall without the extreme boundary line, which circumscribes the class, and are carried, by new affinities, within the limits of some other family. The affinities, which determine these arrangements — constituting the attractive principle, in virtue of which, the individual members of each group find their appropriate positions — will consist in *all* the phenomena and relationships of the several diseases, and not in any limited and arbitrarily chosen portion of them; those which are the most constant, characteristic, and essential, exerting the strongest power. Each class, or family, thus constituted, will be natural and perfect, just in proportion to the number and importance of the affinities, which bind its several members together.[8]

One of the best and purest examples, one of the most perfect models, of such a family, is to be found in the exanthematous fevers. They constitute what may be called the *type-family*, amongst these groups. Occupying the central region of this group, we find small-pox, cow-pox, chicken-pox, measles, and scarlet fever, bound

---

[b] See Whewell's Philosophy of the Inductive Sciences, vol. i. p. 476, 477.

closely together, by numerous and very in-[265]timate affinities. They are all marked by the presence of that general morbid condition, designated by the term *fever*; they are characterized, each of them, by the presence of a peculiar cutaneous eruption; they are all self-limited, in duration; they pass through a regular series of processes, or changes, constituting so many distinct periods, or stages; this limited duration, and these several processes, and stages, cannot be much modified, or interfered with, by art; each of these diseases is capable of propagating itself, by means of a specific poison, or contagious principle; and, finally, they rarely affect the system more than once. At distances, farther and farther removed from this central position, we shall find the disease called *roseola*; nettle rash;[9] erysipelas; plague; malignant pustule,[10] and, perhaps, some others. These latter possess several of the characters, which belong to the former, but not all of them; and as the affinities between them and the type-species, become fewer and feebler, they gradually recede from the central region which these occupy.

Another example of this family arrangement is to be found in those diseases called, simply, *fevers*. Amongst these are intermittent, remittent, and congestive fevers; typhoid and typhus fevers; and perhaps some others. The central point of this group, or the circle nearest to it, is occupied by intermittent, remittent, and congestive fevers, all which may properly be considered as forms, or [266] varieties, merely, of a single disease. Without these, and farther removed from the central point, are placed typhoid and typhus fevers, whether these are regarded as essentially distinct and separate diseases, or only as forms of the same disease. The group, thus constituted, is less natural and perfect, than the former; but its members seem to have more affinities amongst themselves, than they have for any other diseases; although it may be a question, whether typhus and typhoid fevers, placed in the outer limits of this family, do not more properly belong to the preceding group of febrile exanthems.[11] The members of both these classes are, very frequently, at least, marked by an alteration of the blood, consisting in a diminution in the relative quantity of its fibrine.

A third great family is formed by the local phlegmasiæ. Their common character is to be found in the circumstance, that they all consist in that peculiar and well known morbid process, called *inflammation*; and further, that they are, also, generally associated with fever. The latter condition, however, differs from the similar condition in the two preceding families in this, — that it depends directly and immediately upon the local inflammation, as its cause, — goes along with it, and subsides with its disappearance. This family is more numerous and complicated, than either of the two former; but the affinities between its members are, nevertheless, many and strong. It [267] is divisible into several subordinate groups, depending upon secondary, or subordinate, affinities. One of these latter consists of inflammations, seated in the serous membranes; another of inflammations, seated in the mucous membranes; a third of inflammations, seated in the parenchymatous structure of the organs; and still other like subdivisions depend upon certain modifications in the character of the inflammation itself. A peculiar condition of the blood, consisting, principally, in an increase in the relative proportion of its fibrinous element, constitutes a striking and important affinity between the members of this extensive family.

There is another natural assemblage of diseases, marked by characteristics widely different from those of either of the preceding classes. They have been called the *neuroses*. The *type-species* of this class are *tetanus*, and *epilepsy*. Then, at various distances from these, and grouped about them, are chorea, hydrophobia, paralysis agitans,[12] some other varieties of paralysis, delirium tremens, hysteria, catalepsy, and so on. The most striking affinities, which bind these diseases together, are the following, — deranged and irregular action of the voluntary muscles, or entire loss of power over them; absence of fever; and absence, also of any constant and appreciable alteration, either of the solids, or fluids, of the body. This class is a very natural and striking one.[13]

Another example of one of these families is to [268] be found in the several forms of carcinoma, tubercle, and their allied affections. The type-species of this class, are the above-mentioned diseases; they occupy the central point of the group. They are general, or constitutional, maladies; accompanied and characterized by certain local lesions, or morbid depositions, of a strongly marked and peculiar nature. These local lesions and depositions have a tendency to repeat and multiply themselves, in different parts of the same or of several organs; and this tendency, in many cases, is quite beyond the control of art. United to these type-species, by similar, but looser, affinities, are the several forms of scrofula, and syphilis.

It is not necessary to extend this kind of illustration any further. As I have just said, I am not engaged in the formal task of framing a methodical nosology. My object is merely this, — to endeavor to exhibit the true principles, and the essential conditions, of such a nosology; and to indicate the process, by which its construction is to be accomplished. I will only observe, further, that the family groups themselves, into which, individual diseases, according to these principles, and by these methods, have been distributed, are not susceptible of arrangement, under any two or more great, primary classes, like the vasculares and cellulares, the endogens and exogens, in the vegetable kingdom; but that they may, in their turn, like the individuals of which they are composed, be naturally grouped together, or disposed [269] in each other's neighborhood, according to the general similarities between them; and that whatever individual diseases there may be, the characters of which are so obscurely marked, as to render their true position doubtful; or the affinities of which for other diseases are so feeble, as not to bring them within the extreme boundary line of any one of these natural families, must be suffered to remain, provisionally, at least, in the unoccupied spaces between such of the groups, as they are least unlike.

This kind of classification, it will be readily seen, differs from that in many of the positive sciences in this, that it is *approximative*, and not *absolute*; the members of the several classes being held together only by general affinities, in some cases numerous and intimate, and in others few and remote; instead of being immovably fixed in their places, by the presence of some one, or more, invariable, and absolutely identical properties. This peculiarity in the character of these family groups of the objects of natural history has been very clearly and comprehensively stated by Professor Whewell. He says, — "Though in a natural group of objects, a definition can no longer be of any use as a regulative principle, classes are not, therefore left quite loose, without any certain standard or guide. The class is steadily fixed, though not precisely limited; it is given, though not circumscribed; it is determined, not by a boundary line without, but by a central [270] point within; not by what it strictly excludes, but by what it eminently includes; by an example, not by a precept; in short, instead of a definition, we have a type for our director. A type is an example of any class, for instance; a species of a genus, which is considered as eminently possessing the characters of the class. All the species which have a greater affinity with the type-species than with any other, form the genus, and are ranged about it, deviating from it in various directions and different degrees. Thus a genus may consist of several species which approach very near the type, and of which the claim to a place with it is obvious; while there may be other species which straggle farther from this central knot, and which yet are clearly more connected with it than with any other. And even if there should be some species, of which the place is dubious, and which appear to be equally bound by two generic types, it is easily seen that this would not destroy the reality of the generic groups, any more than the scattered trees of the intervening plain prevent our speaking intelligibly of the distinct forests of two separate hills.[c] [271]

Nothing can be truer, than the foregoing representation; and however much we may be disposed to complain of the want of entire absoluteness, and apparent simplicity, in the principles and the results, of this natural classification, we are to remember that this want constitutes an inherent element in all the phenomena and relationships of natural science. In connexion with the particular subject before us, — the methodical arrangement of diseases, — we are to remember that classes, or groups, or families, cannot be created, arbitrarily, and at will, by our own skill and ingenuity; and so squared and adjusted to each other, as to conform to our preconceived plans of artificial simplicity and order. We must take the objects of this arrangement, — individual diseases, — as we find them, — as they exist in nature, — with all their imperfections on their heads; and that arrangement, and that only, is the true, philosophical, and natural one, which recognizes their real character, and is founded upon *all* their similarities and dissimilarities; their differences and affinities.

In regard to the subject of definitions, I will merely say; that although an adequate definition of a group, or family, may be given in a few words, this is not often the case with a species, or a single member, of one of these families. Such a definition must include *all the important and* [272] *more constant phenomena and*

---

[c] Phil. of Ind. Sci. Vol. i, p. 476, 477.

"Diseases, dissimilar, having a symptom in common, as for instance, a cough. There are two pictures, each with a house in it, but one with trees, cattle and a river; the other with carriages and human figures. You may as well swear that the one and the other are alike, because they have the house in common. My good madam, by sticking to the cough as evidence of identity, you reason not a whit better than good master Fluellen, when he found an M, both in Macedon and Monmouth." *Dr. Beddoes's Common Place Book.*

relationships of the disease; it must be a comprehensive and clear enumeration of its elements. The omission of any one of these elements renders the definition, so far, inadequate and defective. And it is no reply to this, to say, that such a definition is only a description of the disease; and that an essential condition of a definition is extreme brevity. This condition is an arbitrary one; and any definition is inadequate and defective, unless it does really define, or describe, the disease. It should be a compact, methodical, and summary description.

CHAPTER XV.

Relations of Vital and Chemical Forces.

I MIGHT very properly have finished this essay, with the termination of the last chapter. I have treated, I think, and sufficiently in detail to answer my purposes, all the subjects, which go to constitute the essential body of medical philosophy. But, there are certain other matters, which, if they do not form a necessary part of this philosophy, are still nearly enough related to it, to justify me in taking some brief notice of them here.

Amongst the subjects, to which I allude, one of [273] the most interesting is that, which refers to the true powers, and relations amongst each other, of the *Mechanical*, the *Chemical*, and the *Vital Forces*, in the production of the aggregate phenomenon of life. The question, as to the proper adjustment of these several powers, or forces, and of the subordination of one to another, which really exists — so far, especially, as regards the two latter of the three — has always excited a good deal of attention; but an unusual interest has been recently given to it, by the publication of the researches and opinions of Leibig.[1] Instead of venturing upon a full and systematic discussion of this subject, I propose merely to make two or three such remarks, as most naturally and obviously suggest themselves.

The first of these remarks is this — that as an essential prerequisite to the formation of any distinct and positive ideas upon this subject — we must define, as clearly and absolutely as possible, the limits of the meaning which we attach to the words *chemistry*, and *chemical force*, or *affinity*. And in endeavoring to do this, it seems to me, that there is but one course of procedure, and that a very plain one. The term *chemistry* must be held strictly to its legitimate, and generally recognized, signification; it must be rigorously limited, in its application. If we make use of it, with the qualifying prefixes, — *modified*, and *vital*, — to designate actions and processes, with which, without this qualification, we admit that it has no-[274]thing to do, we confound all philosophical distinctions, destroy the intelligibleness of our language, and, by an assumption which we have no authority for making, beg the very question at issue.

In the second place, after having thus fixed the limits of the meaning, attached to the term, *chemistry*, we have no right to deny its agency, in the accomplishment of any transformations, or processes, merely because these take place *within the living body*. If chemistry can separate oxygen from carbon, and unite them again, to form carbonic acid, *without the body*, let it assert and enjoy the power of doing the same thing *within it*. If it can combine, in the laboratory, the elements of formic and of oxalic acid, so as to produce these substances, certainly, we are not justified in denying to it the power, and the sole power, in bringing about the same union and transformation in the living animal and the living plant. If the production of animal heat can be fairly and fully explained, and accounted for, on chemical principles, there is no reason why this should not be done. If the chemical action, between the elements of the food, and the oxygen of the air, is amply sufficient, as Leibig asserts, to explain all the phenomena of animal heat; and if this explanation is in no way inconsistent with other phenomena, or contradicted by other equally sufficient or adequate explanations, it is, certainly, perfectly logical and philosophical, to adopt it as the true [275] one. And so in regard to all the processes and functions of the living organization. As far as these consist in chemical actions and chemical changes, let them be attributed to such actions, and such changes. Let chemistry push her researches into the remotest accessible recesses of the living economy, and let her claim, for her own, every process, every act, every transformation, over which she can establish a legitimate jurisdiction. If she can follow the simple substances — the carbon, the oxygen, the hydrogen, the nitrogen, and so on — which constitute the principal ingredients of all nutritive material, in their various combinations — in their compositions and decompositions; — if she can track their mazy wanderings, through the labyrinth of the organic economy — if she can determine the parts which they severally play while in it, and the methods and avenues, by which, and through which, they are finally disposed of, after the purposes of their introduction have been accomplished, why should we hesitate to let her do so? This is her true mission; and in thus fulfilling it, she invades no region to which she has not a clear title — she usurps no power that does not rightly belong to her. But when chemistry attempts to go further than this, she should be rebuked, and her pretensions should be resisted. It would be extraordinary, indeed, in the words of Leibig, "If this *vital principle*, which uses everything for its own purposes, had allotted no [276] share to chemical forces, which stand so fully at its disposal."[a] But let us add, in the language of the same writer, — "A rational physiology cannot be founded on mere reactions, and the living body cannot be viewed as a chemical laboratory."[b] And again, — "We shall obtain that which is attainable, in a rational inquiry into nature, if we separate the actions belonging to chemical powers from those which are subordinate to other influences."[c] In the words of Professor Whewell, "Life is not a collection of forces, or polarities, or affinities, such as any of

[a] Leibig's Organic Chemistry. Boston ed. p. 115.
[b] Ibid. Preface, p. xxx.
[c] Ibid. p. 115.

the physical or chemical sciences contemplate; it has powers of its own, which often supersede those subordinate relations; and in the cases where men have traced such agents in the animal frame, they have always seen, and usually acknowledged, that these agents were ministerial to higher agency, more difficult to trace than these, but more truly the cause of the phenomena.[d] Chemistry is subordinate to Life; the former is only the handmaid of the latter; — a most "tricksy spirit," indeed — a "delicate Ariel" — but still subject, like this creature of the poet's fancy, with other and "meaner ministers," to a more potent magician; obeying his behests, and doing his biddings. Into the higher processes and functions of life, there is no evi-[277]dence, whatever, that she is admitted even, either as agent or co-worker; these processes and functions, so far as our present knowledge enables us to determine, have no resemblance to her operations; it is not only that the former *transcend* the latter — the two are *wholly dissimilar* in their *nature*, and to refer them to the same class of agencies is to destroy all distinctions, and to confound all logic and philosophy.

Leibig proposes what he calls a *theory of fever*; a theory of disease; and theories, also, of the actions of remedial substances. But all these theories are manifestly partial, and incomplete. They embrace only the *chemical element* of pathology and therapeutics. It would be a monstrous perversion of the truth, and a wild departure form all sound philosophy, to receive them as adequate and full theories, or interpretations. They contain, as I have just said, only one element, and this element, secondary and subordinate. *The chemico-anatomical relations, of widely different substances, of the materia medica, may be nearly or quite identical; and the same chemical theory may be applicable to diseases, essentially unlike each other, in all their most important phenomena and relationships.*

And this leads me to the expression of an opinion, with which I will conclude the present chapter. Leibig says, — "The most beautiful and elevated problem for the human intellect, is the discovery of the laws of vitality." Professor [278] Whewell says — "In order to obtain a science of Biology, we must analyze the idea of Life;" and again — "In physiology, what a vast advance would that philosopher make, who should establish a precise, tenable, and consistent conception of Life."[e] By the several terms — *laws of vitality* — *idea of life* — and *conception of life* — used by these distinguished philosophers, must be meant, I suppose — so far as any definite and intelligible meaning can be attached to the terms — not such laws as are ascertainable by common observation, or such ideas and conceptions, as result from this observation — but laws and ideas of a more hidden and subtle, but positive and elementary, character. The feeling, which is expressed in this and similar phraseology, has been very common, especially in the minds of men mostly occupied in the cultivation of the exact sciences. I have alluded to this feeling before. It involves the idea, and leads to the belief, that the science of life, in all its manifold relations, is to be rendered complete and positive, only by the discovery of

[d] Hist. Ind. Sci. vol. ii. p. 403.
[e] Phil. Ind. Sci. vol. ii. p. 122.

its ultimate conditions and phenomena — the primary actions and processes, from which all the subsequent and more obvious phenomena are supposed to flow; — by the detection and establishment of some original and fundamental relationship — like that of gravitation — which shall not only reveal and render intelligible all the mysteries of organic life, but [279] which shall also include within itself the reason and the laws of *all* its phenomena, processes, and relations. Now, it seems to me exceedingly doubtful, to say the least of it, whether there is any rational foundation for this opinion, or feeling. Certainly, no efforts to penetrate the secrets of organic life, and to reach its first and fundamental conditions and actions, should be discouraged; it is never possible to say, before hand, to what results the discovery of any truth may lead; but it seems to me, that the character of the phenomena and relationships of life, so far as we understand, or are able to comprehend it, does not justify the extravagant expectations to which I have alluded.[2] And in reference, especially, to the great practical departments of medical science, although some benefit may yet accrue to them, from minute and recondite chemico-pathological, and chemico-therapeutical researches, I cannot hesitate in expressing the conviction, that our most valuable and available knowledge is still to be derived from the study and observation of *the more obvious and manifest phenomena and relationships, in which these departments consist*. The delicate analyses of the chemist may show, for instance, that a certain class of alkaloids — are chemically related to certain organs and tissues of the body — the nervous — which organs and tissues they affect, in a special and peculiar manner;[f] and this rela-[280]tionship becomes a new ingredient in our knowledge — a positive addition to the science of life; — but it is very questionable, after all, whether practical medicine will derive any great benefits from the discovery. I cannot yet see any good reason to believe, that an acquaintance with the chemical relations of opium and the brain, will shed any very clear light upon the remedial properties of this substance, or furnish us with any new or valuable guides, in its application to the cure of disease. And similar remarks may be made, in regard to the theories of disease, resulting from their chemical relationships alone. A knowledge of the chemical elements, merely of that morbid condition, called *fever*, can hardly be expected to lead to any very important results, so far as its treatment is concerned.[3]

But, be this as it may, it is still true, that no conceivable perfection of our knowledge can in any way alter the essential nature of medical science, or change the modes and processes by which the science is cultivated. No matter how high and complete this knowledge may hereafter become, — no matter what discoveries may be made, — no matter how far behind, our present rude, and superficial, and imperfect attainments may be left, — no matter how refined, and subtle, and absolute, our insight may sometime become, into the ultimate processes of the living economy, — no matter what new and simple laws may be discovered, — still, under these and under all pos-[281]sible conditions, will the character of the science remain unchanged. The materials of all our knowledge will consist of appreciable

---

[f] Leibig's Organic Chemistry, p. 185.

phenomena and their relationships, ascertained by observation; and the connexion of these phenomena, and the nature of these relationships, will still continue to be, just as they now are, matters for interpretation, for speculation, for theory; and the science will still remain where it ever has been, where it now is, — in the former, and not in the latter.

CHAPTER XVI.

Future prospects of medical science. Conclusion. Causes of the slow progress and imperfect state of medical science. Diagnosis must precede therapeutics. Reasons of this. Complexity of therapeutical relationships. Probable extent of our power over disease. French medical observation. British medical observation. American medical observation.

THE slow progress and the present imperfect state of medical science, when compared with most of the other natural sciences, is attributable to two causes. One of these causes is to be found in the nature and character of the science itself, — in the almost infinite variety, extent, and complexity of its phenomena and their relationships. This cause is inherent and irremovable, and it is, in itself, sufficient to have kept the science behind the others to which I have referred. [282] The second cause is to be found in the general misapprehension, which has always existed, in regard to the true nature and objects of medical science, and the best and, indeed, the only methods of promoting its progress. The latter cause has been vastly more instrumental in retarding its advancement, than the former. This cause is not inherent, and may be easily removed. It is, even now, gradually becoming feebler and less extensive in its influences; and in some regions of the general domain of our science it has almost wholly disappeared. An endeavor to estimate the probable result of its entire disappearance, and the substitution in its place of a true philosophy, and of the only legitimate and productive methods and processes of investigation, will form an appropriate conclusion of our labors. Let us consult the signs in our zodiac, and see how far we can cast the horoscope of the destiny which awaits us.

The history of practical medicine, especially, during the last twenty-five years, and a right appreciation of its character, and the conditions and means of its progress, furnish us with very positive assurance, that many of its most important laws will gradually, but steadily and certainly, be carried forwards to their entire and final establishment. The foundations of many of these laws, — and of those too most difficult of determination, — have been already broadly and securely laid; and although many years must elapse, amidst earnest, unremitting, and conscientious toil, be-[283]fore these laws can be *definitively* and *fully* settled, it is not possible, in the nature of things, that we can be deceived, or disappointed, in this consummation, so devoutly to be wished. The minute and thorough study of diseases, in all their aspects, phases, and relationships, which is now prosecuted, with so much zeal and fidelity, cannot fail of leading to the result of which I have spoken. The great laws of pathology and its relations, — of etiology, and therapeutics, — are sure to be ascertained; each successive year will add something to their development, in the steady accumulation of legitimate and authentic materials, and in their disposition and analysis, so that, in the end, the *entire natural history of diseases* will be made out and written.

In this progress of medical science, which we thus confidently anticipate, some of its branches will take precedence of others. Diagnosis, for instance, will be in advance of therapeutics; and this for two reasons. In the first place, the elements of the former are fewer, and less complex in their relationships, than those of the latter; and in the second place, diagnosis is an *essential prerequisite* of therapeutics. These are amongst the reasons why improvements in the treatment of disease, especially for the last twenty-five years, have not kept pace with the advances, which have been made in our knowledge of disease itself. After our knowledge of pathology, and our nosological diagnosis growing out of this, have reached [284] their highest attainable point of accuracy and positiveness, there is still left an almost interminable field of investigation, in the study of the relationships between the morbid condition, thus ascertained, and the substances and agencies in nature, which can in any way affect or influence this condition. Let us look, for a single moment, at the extent and the complexity of these relationships. They are almost infinite. Look at any single disease, even of the simplest and best settled character; and let us suppose that all its elements, as far as this is possible, in the nature of things, have been accurately ascertained. Before our therapeutical knowledge of this disease can be said, in literal strictness, to be *complete*, we must know the effects and influences, which *all the substances and agencies in nature are capable of producing upon it*; and we can know this only by direct observation of the effects themselves. We must know how it will be modified by each and all of the different vegetable productions of the earth; by each and all of the mineral substances, in their manifold forms of chemical combination; by changes of temperature, and other meteorological conditions; by light, by electricity; by food; by drink; by exercise; by the state of the mind, and so on.[a] Now, when it is remem-[285]bered, that these substances and

[a] The doctrine, thus stated, sanctions the constant introduction and trial of new remedies; since until any given substance is tried we do not and cannot known what properties of a remedial nature it may be endowed with. *All* substances, in their remedial characters, were once new; calomel, antimony, opium, Peruvian bark, were once, and some of them not very long ago, new remedies, and any philosophy that would reject the trial of a remedy now, *because* it is new, would of course have rejected the trial of these on the same ground. But, let me say, there is no man, anywhere, who regrets, more sincerely, than I do, the multiplication which is constantly taking place of the so called articles of the materia medica. There is probably no man more entirely skeptical in regard to their alleged properties and virtues, than I am. There is no man who has been in the habit of using a smaller number of them. There is nothing in the whole range of medical history, which shows so miserable a logic, and so false a philosophy, as the introduction of this multitudinous assemblage of new remedies, *with the properties which are so confidently assigned to them.* But then the fault and the error are, — not that the remedies are new, — but

agencies are, many of them, acting together, — that it is exceedingly difficult, in many cases, to separate the influence of one from that of another, in our own endeavors to estimate the real agency of each; and, furthermore, that the elements of the disease itself, so far at least as its therapeutical relationships are concerned, are more or less fluctuating and changeable, — it must at once be seen how [286] true it is, as I have already said, that positive therapeutical knowledge is more difficult of attainment, than any other in the entire circle of medical science.

But, notwithstanding all these formidable and inherent difficulties, this knowledge has made, within the period of which I am speaking, great and positive advances. The effects of many remedies are much better understood, and their value much more accurately appreciated, than formerly. And I believe, that hereafter, this department of our science and art is destined to a more rapid and positive advancement, when compared with the other departments, than has hitherto been its lot. The first essential condition of this advancement, — the accurate and positive diagnosis of disease, — has to a good degree been fulfilled. The first element in the problem to be solved has been ascertained; and we accordingly find, that the attention of many of the best minds in the profession is not turning in this direction. This is the natural course of events. The seat, the character, the regular march, and the tendencies, of the disease, having been first ascertained, the next thing to be done is to find out the best methods of preventing, of modifying, and of curing it. This is what many of the great pathologists of the present day are actively and zealously engaged in endeavoring to do. This is the great mission which now lies immediately before us; this is to constitute the great work of the next and succeeding generations. [287]

I should be doing great injustice to my subject, if I did not mention, as prominent amongst the therapeutical improvements of the last quarter of a century, the change which has been gradually taking place, in the use of violent and dangerous remedies. I am inclined to regard this change as one of the greatest blessings, which modern medical observation has conferred upon the human race; and it is but fair to admit, that absurd as the *system* of homœopathy is, and unsupported as its pretensions are, so far as its *peculiar treatment* of disease is concerned; it has, nevertheless, done great good *by its practice*, — its scrupulous adherence to a strict regimen, and its avoidance of all injurious remedies, — in the furtherance of this revolution.[b] The

---

that the evidence of their value and efficacy is so utterly wanting. My own opinion is, — an opinion founded upon the history and experience of all the past, — that the number of substances endowed with active, and peculiar or characteristic, remedial properties, is small. But whether this number is small, or large, can be determined only by observation and experience, or *trial*. The true course of the philosophical physician is, — not to reject the medicine *because it is new*, but for the reason, abundantly sufficient in regard to nineteen twentieths of the articles of the official pharmacopœias, — that there is no satisfactory evidence that it is worth anything, and one of the most certain and beneficial results of a correct medical philosophy will be the final expulsion and banishment of these aliens and imposters from the domain of our science.

[b] "It has been sarcastically said, that there is a wide difference between a good physician and a bad one, but a small difference between a good physician and no physician at all; by which it is meant to insinuate,

conviction has been steadily gaining ground, and spreading itself abroad in the medical community, not only that *heroic* remedies, as they are called, are often productive of great mischief, and should never be lightly or questionably used; but that in very many cases of disease, *all medicines*, using this word in its common signification, are evils; and that they may be dispensed with, not merely with negative safety, but to the actual benefit of the subjects. The golden [288] axiom of Chomel, that it is only the *second* law of therapeutics *to do good*, its *first* law being this — *not to do harm* — is gradually finding its way into the medical mind, preventing and incalculable amount of positive ill. The real agency of art is more generally appreciated than formerly; and its arrogant pretensions much more truly estimated and understood. It is coming every day to be more clearly seen, that perhaps its most universal and beneficent function consists in the removal and avoidance of those agents, the action of which is to occasion or to aggravate disease; thus giving the recuperative energies of the system their fullest scope and action, and trusting to them, when thus unembarrassed and free, for the *cure* of the disease.[c]

It is melancholy to think what an enormous aggregate of suffering and calamity has been occasioned by a disregard of the axiom which I have quoted. Our means for the direct removal of disease are limited in extent, but it is not so with our power to augment and to cause it; this is unlimited. Difficult as it may be to cure, it is always easy to poison and to kill. We may well congratulate ourselves and society, that the pri-[289]mary and fundamental truths, of which I have been speaking, are finding their right position, and producing their legitimate results; and that long abused humanity is likely, at no very remote period, to be finally delivered from the abominable atrocities of wholesale and indiscriminate *drugging*.

I cannot forbear remarking, by way of parenthesis, that this evil, in addition to the many others, which I have already had occasion to enumerate, has been greatly aggravated, and in many instances wholly produced, by the influence of *à priori* medical doctrines. The whole history of medicine will show that the most flagrant abuses of this character have always been the direct results of these mischievous and malign influences. I have seen a patient with a cold sweat of absolute agony on his forehead, occasioned by the application of a painful surgical remedy, the use of which was *deduced* exclusively from the loosest *à priori* physiological considerations, wholly unjustified and unsupported, either by common observation, or common sense. Very fortunately for its preservation and welfare, the human system has been endowed with wonderful powers of resistance to unfriendly and pathological influences; and although in this and in other similar cases, the innocent victims to *rational* physic ! may have escaped with their lives not seriously endangered, and their future health not gravely impaired, one would think that an

---

that the mischievous officiousness of art does commonly more than counterbalance any benefit derivable from it." *Sir Gilbert Blane.*

[c] "This, I apprehend, is so well understood among well educated physicians, that the word *cure*, as applied to themselves, is proscribed as presumptuous, and rarely, I believe, escapes the lips of any practitioner, whose mind is duly tinctured with that ingenuous modesty which characterizes the liberal and correct members of the profession." *Sir Gilbert Blane.*

art, which claims the right to a *divine* [290] appellation, and one great purpose of which is the prevention and removal of physical suffering, ought at least to be careful how it thus indulges in the perpetuation of gratuitous cruelty.

There is one question, that very naturally suggests itself in connexion with the subject before us, which is this; — what are the prospects and probabilities in regard to *the real extent and degree of our power over disease*; and how far is this power likely hereafter to be carried? To this question, it is not possible to return anything like a positive answer. It is a favorite doctrine with many, that *all disease* is finally to be brought under the control of art; — that there is no malady, to which the human body is liable, for which either a preventive or a cure does not lie somewhere concealed in the unexplored or undiscovered recesses of nature. Dr. Rush pleased himself with the idea, that some healing plant might be blossoming in our own valleys, endowed with the property of curing consumption. The whole tone and tenor of this essay must be proof enough, that my own hope of the future is strong and bright; but let us be careful how we mistake for the rational indications and the sober teachings of philosophy, the golden day-dreams and the fairy imaginations of a speculative optimism. There are diseases, over which, after the long trials of many centuries, we have failed to obtain any power. Modern skill, notwithstanding its confident promises, and its high pretensions, has not, to [291] any appreciable extent, diminished the rate of mortality from consumption; and the family of cancerous diseases, pursue now, as they did in the days of the Egyptian priesthood, wholly regardless of our interference, their painful and uninterrupted career. Some of the most terrible amongst acute affections, — hydrophobia and tetanus, for instance, and the graver forms of many epidemic diseases, — cling to their victims with a tenacity, which no strength or cunning of our can unfasten or relax.

It ought to be remembered, then, when our art is reproached, as it so often is, with *impotence* as well as *blindness*, that the reason of this impotence may sometimes be found *in the constitution of nature*, and not in any fault of its own. For anything that yet appears to the contrary, there are many morbid conditions, which art has no power to remove, — the means for the removal of which, I mean to say, *do not exist*. It is not, merely, that we have failed to discover them; — it may be, that they *are not in nature*. These morbid conditions may absolutely have no therapeutical or curative relationships.

Precisely similar considerations are applicable to many of the *etiological relations* of disease. I can see but small reason to believe, that the mysterious and overwhelming energies, constituting the causes of such epidemics, as the black death, the Asiatic cholera, the typhoid pneumonia and spotted fever of New England,[1] will ever be coun-[292]teracted, neutralized, or destroyed, by the skill or achievement of human science. But in regard to some of the most common and destructive diseases of a chronic character, we have better grounds of hope. Somebody has said, that acute diseases are amongst the direct chastisements of Providence; while chronic affections are the natural fruits of our own disobedience to the laws of health; and the idea contained in the remark is not wholly without foundation. The causes of many of these diseases are obviously to be found in violations of the known conditions of health; and in all such cases the power of avoiding the diseases is placed in our own hands.[2] I have small faith in the discovery, or the existence, of any specific antidote, or cure, for consumption; — I have but a feeble hope, that any balm will ever be gathered from the green and blossoming surface of the earth, or dug from its bosom, or distilled from its enveloping atmosphere, so sovereign and potent, as to arrest the deposition of tubercle, or to expel it from the body; but I do cherish the belief, as well as the hope, that by a general and strict conformity to the ascertainable conditions of health, aided by that stern but beneficent arrangement of nature, according to which, she issues her inexorable edict of extermination against the race, which has become deteriorated by flagrant and long-continued disregard of these conditions, this disease may yet be staid in its hitherto resistless career, and shorn of its terrible strength. [293]

By whom the future triumphs of our art are to be achieved; — by whom the great movement, which the last twenty-five years have impressed upon medical science; is to be sustained and carried on; and to whom the honors and rewards of sustaining this movement, and of urging it forward to the development of its full results are to accrue, — can hardly be regarded as questions of vain curiosity, merely, or of doubtful speculation. A careful study of the tendencies of the general medical mind, in the different leading portions of the scientific world, and an impartial estimate of its character, and peculiarities, would not fail, I think, of furnishing answers to these questions, at least of probable or approximative certainty. Such a comprehensive study, and estimate, my own want of familiarity with the languages, the literature, and the science of all the countries of continental Europe, excepting France, — even if there were no other reasons, — would prevent me from undertaking; and I propose, merely, to take a brief and cursory view of the questions, which I have stated, so far only as they relate to France, Great Britain, and the United States; and, even in this limited view, I shall confine myself entirely to those portions of our science and art consisting of internal, or medical pathology, and its relations, — or, in other words, — to what is commonly called *practical medicine*.

Let us look first at the character of medical [294] science, and of the general medical mind, in France. And the peculiarities of this character it requires but little pains to discover; they are impressed in such broad and deep lines on the history of medicine in France, during the present century, as to force themselves on our notice, and to render this period one of the most remarkable epochs in the annals of medical science. It may be designated as that of the origin and establishment of the *Modern School of Medical Observation*. This school is characterized by its strict adherence to the study and analysis of morbid phenomena and their relationships; by the accuracy, the positiveness, and the minute detail, which it has carried into this study and analysis; and by its rejection, as an essential or legitimate element of science, of all *à priori* reasoning or speculation. The spirit which animates, and guides, and moves it, is expressed in the saying of Rousseau, — *that all science is in the facts or phenomena of nature and their relationships, and not in the mind of man, which discovers and interprets them*. It is the true *protestant* school of medicine. It either rejects as apocryphal, or holds as of no binding authority, all the traditions of the fathers, — unless they are sustained and sanctioned, by its own experience.[d] It [295] appeals in all things directly to nature, and it asks, — not, what *may be*? or what *ought to be*? but, what *is*? — not, *how* things are? or *why* they are? but again, *what* they are? Holding that medical, as well as all other, science, should have but one aim and object, — *to ascertain the actual constitution of things*, — it professes an entire scientific indifference, as to the issue and result of its researches, provided, only, that this issue and result approach, in the nearest possible degree, to the absolute truth; and it adopts and pursues what it conceives to be the only method and means of accomplishing this end.

This school has given birth in France to a series of very remarkable works, — the fruits and records of its labors, — a summary notice of enumeration of the principal of which will serve better, perhaps, than anything else, to illustrate [296] its character and its achievements. One of the earliest formal manifestations of that spirit and tendency of the French medical mind, which led, in their farther progress and their full development, to the formation of the school, of which I am speaking, is to be found in the work of Prost,[3] entitled, "*Medicine illustrated by observation, and the examination of bodies*;" published in 1804. This publication, although in many respects faulty and imperfect, marred by *à priori* reasonings and gratuitous conclusions, is conceived, and executed, on the whole, in the right spirit; and marks very clearly the bright dawn of the new era. It is mostly made up of short histories, — evidently fair, careful, and trustworthy, — including the symptoms, and the lesions found on examination after death, — of more than one hundred cases of various kinds of disease. Four years later, in 1808, appeared Broussais's immortal *History of Chronic Inflammations*; a work which constituted one of the corner stones of that temple of true science, which he himself, at a later day, and under the influence of a false philosophy, strove, with such blind but giant efforts, to destroy.

These works have been followed, in rapid succession, by others of a similar character, covering almost the entire domain of pathology and therapeutics. Prominent amongst these may be mentioned the researches of Corvisart,[4] Laennec,[5] and Bouillaud[6] upon the pathology, diagnosis and treatment of diseases of the lungs and heart; those of [297] Petit and Serres[7] on fever; of Rostan,[8] Rochoux,[9]

---

[d] Louis has been very harshly censured by some of his critics, and especially by a voluminous commentator of our own country, for his low estimate of the actual and positive value of a considerable portion of past medical observation. This censure, it seems to me, is wholly undeserved. It is not possible for any one, who rightly estimates the real nature of medical science, to study at all extensively and carefully its past history without coming to the same conclusion with Louis. Dr. Denman, one of the clearest headed and soundest thinkers among the British medical men, in speaking of those who pinned their faith upon Hippocrates, says, "*they were constantly praising learning at the expense of knowledge*." And in opposition to the commentaries and criticisms, of which I have spoken, hear, further, what Sir Gilbert Blane, — always judicious, always elegant, always free from philosophical fogginess and error, — says upon the same subject: — "*It is a melancholy truth, that there is, perhaps, no department of human knowledge, in which there is so great a want of correctness, with regard to recorded observations, as well as reasoning*. We ought, therefore, to be strongly fenced against the inroads of error in others, as well as in ourselves." *Elements of Medical Logic*, p. 190.

Lallemand,[10] Parent Duchatelet,[11] Martinet,[12] and Durand-Fardel,[13] on disease of the brain; those of Martin Solon[14] and Rayer,[15] on diseases of the kidneys; those of Valleix[16] on neuralgia; of Grisolle[17] on pneumonia; of Rilliet and Barthez,[18] on disease of children; of Andral and Gavarret,[19] on the blood; and those of Chomel, Andral and Louis, on several of the above, and on many other most important and most interesting subjects of pathology, diagnosis, and therapeutics. Since the time of Hippocrates, there has not appeared in any age, or country, a series of contemporaneous publications, upon similar subjects, at all equal to these in extent, variety, and positive value. There is hardly an important point in pathology upon which they have not shed new light, — there is hardly a disease, the diagnosis of which they have not rendered easier and more certain, than it formerly was, while in many cases we are wholly indebted to them for our means of diagnosis; and they have added not a little to the exactness of our knowledge in regard to some therapeutical processes.

The works of Louis were the first mature fruits of what has since been called the *numerical school*, — but which in truth, and strictness, as I have already said, only a numerical method of statement, analysis and appreciation. The opposition with which this method has met; the reluctance with which its value, its importance [298] and its true character have been admitted, even by many men of sound and logical minds, exhibit and illustrate, in a very striking manner, the strong hold upon the general medical mind, and the pernicious influences, of a false and vicious philosophy. One would have supposed, on listening to the furious and blind tirades uttered against the new school, that some Gothic Alaric in medicine had broken into the time-honored and legitimate domains of our science, to despoil it of its choicest treasures, to overturn its venerable altars, to throw down the monuments of its past and present glory, — and everywhere to lay waste and destroy. All true followers of Hippocrates, and of *philosophical* and *rational* medicine, were earnestly called upon to resist the encroachments of this new and formidable invader; and to prevent the corruption of their ancient faith, and the desecration of their ancient shrines, by the inroads of *pathological anatomy, positive diagnosis*, and *arithmetic!* The smoke of this noisy resistance is at length, partially at least, cleared away; and it is clearly enough seen, by every body who will take the trouble to look, that the danger, which excited it, was altogether imaginary.[e] The "numerical method" constitutes no new

[e] In reading the long, captious, and windy *reasoning*, as it is gravely termed, of certain writers, I am sorry to say in our own country, especially, against the authenticity, and the existence, even, of the phenomena and relationships, observed, and ascertained by Louis, we are forcibly reminded of the remarks of Professor Whewell in connexion with a similar subject. "When Galileo," — he says, — "had announced, in 1610, in his *Siderial Messenger*, 'great and very wonderful spectacles,' which he had recently discovered in the heavens, many '*paper philosophers*' of the day, as he termed them, appear to have thought that they could *get rid of these new objects, by writing books against them*." *Phil. Ind. Sci.* Vol. i. p. 363. The books of these "paper philosophers" are now only known through the generous intervention of quotations similar to the foregoing; but the "*great and very wonderful spectacles*," which science and art had just revealed to the "starry Galileo," are still shining, eternal in the heavens; and whatever may be the advances, which medical science shall hereafter make; whatever may be the final issue of its labors, — the observations of Louis upon the natural history of phthisis, yellow fever, typhoid

system of medicine; it has no analogy, whatever, to what have been [299] commonly called *medical doctrines*. As I have already said, it is only a method for the statement, analysis and appreciation of ascertained and positive phenomena and relationships. This is the sum of its character and pretensions. It has no other; it has never claimed any other. It insists upon the use of positive terms and statements, in all cases where positive phenomena and relationships are its objects. It insists upon the necessity, in science, of calling things by their right names; and of stating all facts and relationships, fully, entirely, *rigorously*. It insists upon the necessity of accurately *enumerating* all phenomena that are *numerable*. It alleges, that there are no principles in medicine, — as there are none in any other of the sciences of observation, — which are not the aggregate result, or in other [300] words the simple expression, of facts and their relationships.

> "The very head and front of its offending
> Hath this extent, no more."[20]

This great revolution, the character and tendencies of which I have thus attempted to indicate, has been but very partially felt by the general medical mind of Great Britain. The *British school of observation* may fairly enough be said to have commenced with Sydenham; and its general spirit and principles have continued almost entirely unchanged to the present time. This school has been marked by some of the strongest and best qualities of the British character, — sagacity, shrewdness, and sound common sense. It has been regularly progressive since the time of Sydenham, and it has accumulated a vast amount of most excellent practical knowledge. Its therapeutical resources have been more various and extensive, than those of its continental rival; and if it has done less for the advancement of medicine, *as a science*, it can hardly be doubted, I think, that it has accomplished more, as a useful and beneficent art. Amongst the models and ornaments of this school, may be mentioned Sydenham, Huxham, Cleghorn[21], Heberden,[22] Blane, Pringle, Thomas Percival,[f, 23] John Cheyne,[24] Thomas [301] Bateman,[25] Samuel Black,[26] William Woolcombe,[27] William Brown,[28] Sir Henry Marsh;[29] and many others, their worthy successors, might be added, — the pride and glory of the actual period of British medicine.

The principal defects of the British school are its want of comprehensiveness, of rigorous and positive conclusions, and the habit of mixing up, with its observations,

---

fever, pericarditis, and pulmonary emphysema, will still remain standing, — landmarks of its progress, imperishable as the pyramids.

[f] Dr. Thomas Percival announced as early as 1789, very fully and explicitly, the true pathological character of consumption. "In this malady," — he says, — "inflammation is perhaps only an occasional concomitant; for the tubercles in the cellular substance of the lungs are found to be of a whitish color, and cartilaginous hardness, and to remain solid till they attain a certain size; matter then begins to be formed in their centre; as they grow larger, suppuration advances till they are converted into vomicæ. . . . *Tubercles and vomicæ probably constitute the characteristics of the disorder in every form.*" *Mem. Med. Soc. of London.* Vol. ii. p. 303. The earlier observations of Stark upon this subject, I have not been able to obtain.

reasonings and interpretations altogether hypothetical in their character; and then of regarding these reasonings as more important, more valuable, more essential to the constitution of science, than the observations upon which they are founded.[g] These defects have [302] not however been universal, and the indications are too clear to be mistaken, that they are destined very rapidly to diminish, and finally to disappear altogether. I have had more than one occasion, in the course of the preceding essay, to cite individual manifestations, in British medical literature, of the highest and best spirit of philosophy. I could easily fill many pages with other and similar citations. Even the rigorous numerical method of Louis, although it has been very slowly and reluctantly received into the modern medical mind of Great Britain, was adopted and followed, to a considerable extent, by some of her old observers, with whose names I have already graced these pages. Amongst these I may mention particularly, Dr. Thomas Percival, of Manchester, who exhibits very strongly his fondness for positive numerical data, in a volume of *Medical Essays*, published as long ago as 1776;[30] Dr. William Brown, of Edinburgh;[h] William Woolcombe; John Cheyne, in nearly all his Hospital Reports; and to these may be added, more recently, Dr. James Crau-[303]furd Gregory;[i] Dr. David Cragie;[j] Dr. William Henderson;[k] Dr. John Reid, and Dr. Alexander P. Stewart.[l] Thus, in addition to the more formal recognition of the legitimacy of the claims of the numerical method, by one at least of the leading, and one of the ablest British Reviews,[m] and by some other British authorities of high character,[n] we have the still more conclusive testimony to its

---

[g] "These Essays," says a writer in the *Edinburgh Medical and Surgical Journal*, vol. xxiv. p. 101, "belong to a class of publications for which English physicians have been long eminent. Without forming a complete systematic discussion of any subject, they consist of pathological and practical remarks, *mixed with a good deal of reasoning*, on those subjects with which the line of the author's experience has made him most familiar." In vol. xxxv. of the same Journal there is a capital paper, full of sound philosophy, in the form of a review of *Abercrombie on the Intellectual Powers*; and entitled, *Application of Psychology to Medicine*. The writer says, in reference to the subject before us, — "How often have we occasion to remark, that *matters of opinion* are stated *as matters of fact*; and that an author *instead of limiting himself to a strict and faithful description of what he observes, introduces, apparently unconsciously, his own opinions and inferences*? In such a mind it is impossible to doubt that no distinct line had ever been drawn between *what is observed and actually exists*, and what the observer himself *imagines to exist*."

Dr. David Uwins says, — "To medicine belong philosophical acumen, and *a promptness of drawing correct inferences from occasionally doubtful premises*." *London Lancet*, Feb. 1825.

[h] See a most admirable paper, by Dr. Brown, — "*On the inefficacy of medicine in arresting or shortening continued fever*" — published in *Duncan's Annals*. vol.. vii. 1802.

[i] See a paper by Dr. Gregory, "*On Diseased States of the Kidney, connected during life with Albuminous Urine*," *Edin. Med. and Surg. Journ.* vol. xxxvi. xxxvii., 1832.

[j] Dr. Cragie's *Report of the Cases treated during the Course of Clinical Lectures, delivered at the Royal Infirmary, in the Session*, 1834, 1835.

[k] "*Report on the Epidemic Fever of Edinburgh*." *Edin. Med. and Surg. Jour.* Oct. 1839.

[l] "*Some Considerations on the nature and Pathology of Typhus and Typhoid Fever*, applied to the solution of the question of the identity or non-identity of the two diseases." *Edin. Med. and Surg. Jour.* vol. liv.

[m] See *British and Foreign Medical Review*, July 1841.

[n] Dr. Henry Holland says, — "It must be admitted, however, that the methods of research in medicine at the present time have gained greatly in exactness, and in the just appreciation of facts, upon those of any previous period; — a natural effect of increasing exactness in all other branches of science. A very especial advantage here has been the application of numerical methods and averages to the history of

value, contained in the fact of its practical adoption by the observers above-mentioned. [304] One of the latest and most valuable works on practical medicine, from the British press, — Dr. Robert Collins's *Practical Treatise of Midwifery*,[31] — is arranged and constructed in strict accordance with this method.

The general character of medical science, and the tendencies of the medical mind, in this country, are not marked by any striking peculiarities, and may be very easily stated. They partake of those both of Britain and France, whence, indeed, we have derived them. Up to the period of the publication of Bichat's great work on *General Anatomy*,[32] the medical opinions of the United States were received, almost exclusively, from Great Britain; but few foreign works were generally current amongst us, excepting those of British origin; such of our young men, as went abroad for a medical education, repaired mostly to Edinburgh; and the medical doctrines of North America, so far as she had any, were engrafted on those of Scotland and England. This transfusion of the British medical mind into that of the United States was he natural and necessary consequence of the relations of the two countries; and the dependence of the latter upon the former, although less exclusive than it once was, has never ceased to exist. Nearly all the leading works, in every department of medical science, which have appeared in Great Britain, since the days of Cullen, have been republished, and very extensively read, in this country; while up to the [305] time of Bichat, American publications of French medical works were few and far between. The last fifteen years, however, have witnessed a great change in this respect. While our medical relations with Great Britain are still, as they will always continue to be, numerous and intimate, they are altogether less exclusive than formerly; and they are probably now inferior, in interest, influence, and importance, to those which exist between us and France. Our young men have almost entirely ceased to visit the British capitals, in order to complete their education; and the number of those who have annually repaired to Paris, for this purpose, for many years past, has been very much greater than has ever been the case with Edinburgh and London. The leading works of Louis were earlier and more widely circulated in this country, than in Great Britain; and the principles of his school and method have taken deeper root here, than there. There is now a pretty large and constantly increasing class of young physicians, many of them personal friends and pupils of Louis, scattered through our principal cities, mostly at the North and East, thoroughly imbued with the spirit of their distinguished master, and diligently engaged in the study of positive pathology, diagnosis, and therapeutics.[33]

In looking at the actual additions, which we have made to medical science, although we may fail to discover anything of sufficient importance, [306] very greatly to flatter our national pride, or to strengthen our national complacency — although we can point to no such trophies, as the revolution wrought by Sydenham

---

disease; thereby giving it the same progress and certainty which belongs to statistical inquiry on other subjects. Averages may in some sort be termed the mathematics of medical science. The principle is one singularly effectual in obviating the difficulties of evidence already noticed; and the success with which it has been employed of late, by many eminent observers, affords assurance of the results that may hereafter be expected from this source."

in the treatment of small-pox, Jenner's discovery of the powers of vaccination, and Bright's investigations of certain morbid states of the kidney; — although we can boast of no such achievements in pathology and diagnosis as those of Laennec and Louis; it is still true, that the labors of American medical men have not been altogether barren of valuable and positive results; and when the circumstances of our position, and the general prevalence amongst us of a vicious, or faulty philosophy, are taken into account, we have accomplished as much, perhaps, as could reasonably have been expected. The writings of Caldwell,[34] Rush,[35] and Edward Miller,[36] towards the close of the last, and during the early part of the present, century, did as much, at least, as those of any other authors, in establishing the doctrine of the non-contagiousness of yellow fever. The paper of M. Louis, on emphysema of the lungs, published in the first volume of the *Memoirs of the Medical Society of Observation*, furnishing so large an addition to our knowledge of this disease, is made up of materials, which were the result of the joint labors of its author, and Dr. James Jackson, Jr. — so early lost to his country, and to science.[37] The occurrence of prolonged bronchial expiration, in the first stages of tubercular deposition, and of [307] pneumonia, and its great value as a diagnostic sign, were also first pointed out by Dr. Jackson, Jr. Very important additions have been made to our knowledge of the pathology and diagnosis of tubercular meningitis, and lobular pneumonia, by Dr. Gerhard,[38] of Philadelphia; and the observations of the latter gentleman, in connexion with those of Dr. Pennock, upon the symptoms, causes, lesions, and treatment of typhus fever,[39] constitute some of our most authentic and valuable materials, towards the settlement of one of the most interesting and important questions of pathology — that of the true relations between the two great forms of continued fever. Of a similar character to these, are the observations of Dr. Stewardson,[40] upon the anatomical lesions of bilious remittent fever. Dr. James Jackson,[41] and Dr. E. Hale,[42] both of Boston have added much to the accuracy and positiveness of our knowledge of the common continued fever of the United States; and in connexion with others, have fully demonstrated its complete identity with the typhoid fever of France, and the abdominal typhus of Germany. Acknowledged and positive reforms in therapeutics are of rare occurrence; and I should do great injustice, both to my countrymen, and to my subject, if I failed to notice here the result of American observation, in regard to the treatment of *delirium tremens*. The doctrine upon this subject has been, ever since the time when the attention of the profession was called to the disease, [308] by Dr. Sutton,[43] that the patient *must sleep or die*; and that the only sure means of securing the first alternative, was to be found in adequate, and generally enormous, doses of opium. There were few points in therapeutics which were generally regarded as better settled than this. Dr. John Ware,[44] of Boston, was led, however, more than fifteen years ago, to doubt the legitimacy of this therapeutical dogma; and the results of his own careful observations soon satisfied him, that the sleep of convalescence was much oftener the natural and spontaneous termination of the disease, than the effect of the heroic doses of opium. Subsequent and very extensive experience, in the treatment of this affection, both by himself and others, has fully demonstrated the soundness of this conclusion; and shown very clearly the superior efficacy of other, and safer, modes of management. The therapeutical reform thus commenced has been recently completed, by the remarkable and unparalleled success, which has followed the alcoholic treatment of the disease, by Dr. Gerhard. It ought not to be forgotten, in this connexion, that the profession is indebted to an American physician — Dr. Stearns, of New York — for the revival, and the introduction into general use, in both hemispheres, of the *secale cornutum*, as a special excitant of uterine contractions.[45]

The causes which have led to the differences, that I have thus endeavored to indicate, in the character and results of medical researches, in [309] France, Great Britain and the United States, must continue to operate for a considerable period of time. The means and facilities for the prosecution of these researches are more numerous and available in France, than they are, either in Great Britain, or in this country; the hindering and mystifying influences of a bad philosophy have to a great extent disappeared from the French medical mind, while that of the Anglo Saxon race is yet only struggling out, partially emancipated, from the thraldom of these influences; and, finally, without stopping to inquire whether the spirit of personal, professional jealousy, and of private interest, is as active in Paris, as it is in London and New York, I suppose it as at least safe to say, that there is a stronger and more pervading tone of scientific inquiry, — a higher and wider range of scientific emulation, — in the former city, than in either of the latter. These differences, however, between the several countries will very certainly diminish from year to year; and we have every assurance, which the history of the past, the indications of the present, and the nature of the subject can furnish, that there shall be hereafter, amongst the lovers and seekers of truth, everywhere, a closer and more effective coöperation, than has hitherto existed, in carrying forwards, in their career of illimitable progress and indefinite improvement, all the branches of the great science of life. [310]

INDEX.

Page

Anatomy, definition of 61-64
Sources of our knowledge of 77, 78
Atomic theory 36
American Medical Doctrines . . . 220-246

Brewster, Sir David 23
Births, division of between sexes . . 149-153
Broussais 203-206
Blane, Sir Gilbert 217
Botany, systems of classification in 248-254
British medicine 297-301

Classification, in physical science . . 53-55
of diseases 246-269
Causes, final 79-82
Congestion, a common element . . 127,128
Calculation of probabilities . . . 150-154
Cullen, his theory of fever . . . 183-185
Cooke, Dr. John Esten, his theory . 227-237
his therapeutics 232-235

Chemistry, its functions 269-277

Diamond the, its combustibility . . . 24
Dalton, Dr., his atomic theory 37
Davy, Sir Humphrey 50-52
Definitions 61-67, 268
Diagnosis, reason of its importance . . 117
its relation to therapeutics 117-123
nosological 123-138
therapeutical 139-141
Diseases, their common elements . 126-128
seat of 129, 130

Page

Diseases, nature of 131, 132
Delirium tremens, its treatment . . 304, 305
Doctrines, medical 176-246
general prevalence of . 177-178
their bad influence . . 202-214
American 220-246

Erroneous notions, extent of . 68, 75, 76
Etiology, sources of our knowledge of . 96
not deducible from pathology 97-98
Empirics 100

Fresnel, M. 19-24
Fever, typhoidal, a common element 126, 127
inflammatory, a common element 126, 127
typhoid and typhus, their differences 136-141
Dr. Rush's theory of . . . 221-223
Forces, chemical and vital . . . 269-278
French medical observation . . . 292-297

Gravitation, its nature, or essence . . 42-45
Germination, its phenomena . . . 70-73
Gavarret, M. . 150, 151, 155, 157, 161, 162
Gallup, Dr., his theory 238-240
Gerhard, Dr. 304, 305

Herschel, Dr. 19
Herschel, Sir John 19
Hypotheses, in physical science . . 33-52
how far admissible in medical science . . . 217-219

Homœopathy 185- 197
Holmes, Dr. O. W. 197- 201
Hunter, John 246

Inflammation, sources of our knowledge of 89-93
its therapeutical relations 104-108

Jussieu, his classification 249-254
Jackson, Jr., Dr. 303, 304

Knowledge, its incompleteness . . 32, 33

Law of great numbers 149-151
Linnæus, his classification . . . 248-254
Leibig 271, 274
Louis 294, 295, 296

Marble, its properties and relations . 12-17
Malus, M. 19
Mathematics, functions of 20-22
Methodism 180-183
Matter, constitution of . . 34, 35, 53, 54, 55
Medical Science, propositions in . . 59, 60
its future prospects 278
Miller, Dr. Edward, his theory 226
Miner, Dr., his theory 238-240
Method, numerical . . 175, 294-297, 299

Newton, Sir Isaac 19, 21, 24, 38, 39, 40, 43, 48, 49, 50
Nosologies systematic 246-269
their defects . . . 256-260

Object of essay 5
Observation, the sole foundation of physical science 10-12
the sole foundation of medical science 67-116
its imperfections . . . 163-166
its difficulties 202
Optics, sources of our knowledge of . 18-25
theories of 38-41

Physical science, propositions in . . . 3
nature and elements of 7-9
its laws, or principles 26-32

Physical science, independent of hypotheses 34, 35
Phenomena, invariableness of 27
Physiology, definition of 64
not deducible form anatomy 79-81
Pathology, definition of 65
relations of 66
not deducible from physiology 81-96
qualifications 82-89
its laws 153-155

Philosophy, false, prevalence of . . 67-69
Principles in medical science . . 143-176
how ascertained . . . 158-162
contingent nature of their elements 171-173

Relationships invariable 26-28
Reasoning *à priori*, its functions . . . 24
Respiration, its phenomena 73, 74
Rationalists 100
Remedies, action of on lower animals 115, 116
new 281, 282
Rush, Dr., his theories 221-226

Sexes, proportion of 149-153
Statistics, medical 149-162
Sydenham, his philosophy . . . 210-212
School of observation,
French 292-297
British 297-301
American 301-305

Therapeutics, not deducible from pathology 101-110
not deducible from physiology 111[1]-114
its laws, or principles 155-162
Tully, Dr., his theory 238-240
Thompson, Samuel, his theory . . 240-244

United States, medicine in 301, 305

Wolloston, Dr. 23
Whewell, Professor 36, 267, 268, 273, 275
Ware, Dr. John 305

Young, Dr. 22
[312]

THE PHILOSOPHY OF THERAPEUTICS

The subject of this short essay is the *Philosophy of Therapeutics*. I use this word philosophy here, in the sense that is usually given to it, in similar connexion with other departments of human knowledge and speculation. I propose to consider under this title, the nature and relations of therapeutics; the kind and character of the phenomena which constitute the elements of this branch of science;—the nature, the degree of certainty and positiveness, and the sources, of our knowledge of these phenomena:—and the true methods, and essential conditions of giving to this science that degree of certainty, positiveness, determinateness, and consequent usefulness, of which it is susceptible.

I do not think the importance of this subject can be over-estimated: I think it is almost universally under-estimated. It seems to me that the chief cause and reason of the slow progress, and of the want of certainty, positiveness and completeness of medical science, in most of its departments, and especially in this of therapeutics, are to be found in the general prevalence, always, from the time of Hippocrates to our own day, of vicious and inadequate methods of investigation, growing out of a vicious and inadequate philosophy; and that the first essential condition of the advancement of our science, and especially of therapeutics—of its increased certainty, positiveness, completeness, and consequent usefulness, is to be found in the adoption of legitimate and adequate methods and processes of research and study, founded upon a legitimate and adequate philosophy.

I shall arrange what I have to say under twelve propositions. [1]

Proposition First. The science of therapeutics consists in the phenomena analyzed and classified, which result from the curative relations existing between the various substances and agencies constituting the materia medica, on the one hand, and the morbid actions, tendencies, and conditions of the human economy on the other.[2]

The foregoing proposition consists simply of a definition of the science of therapeutics. It is not an easy matter, in any of the more complex departments of human knowledge, to frame a perfect and unexceptionable definition; and this is especially true of most of the branches of the science of life. That which I have here given is at least sufficiently comprehensive, adequate, and complete, for my present purposes; I suppose no objection will be made to it; and it does not seem to require any commentary or elucidation. I have not included in it, nor added to it, any definition of the art of therapeutics, or of the materia medica. Of the former it is enough to say that that it consists in the practical application of the science. Of the latter, I wish to make one or two remarks, referring especially to its connexions with therapeutics, and its true relations to that science and art. Materia medica and therapeutics are almost always taught in connexion with one another; and it seems to me there is a common feeling that there is a closer and more essential relation between them than really exists. The materia medica consists of all those substances, agencies and influences, which are endowed with the property of removing the diseased conditions, and of arresting, controlling and modifying the disease actions and processes of the human economy. Now what I wish to say is this;—there is no true scientific relation between pure materia medica and therapeutics: therapeutics is not a corollary[3] of materia medica; a knowledge of the materia medica does not include a knowledge of therapeutics; the former is not even a necessary or essential condition of the latter; it is quite separate from it, and independent of it. The natural history of the substance and agencies of the materia medica constitutes one branch of knowledge; their therapeutical actions and relating constitutes another. And these two branches are essentially independent of each other. The natural history of cinchona, its botanical character and relations, its geographical distribution, its anatomy and physiology, its chemical constitution, do not involve in any way its therapeutical properties: they do not even indicate them, or throw any light upon them. Botany and pharmacy do not include therapeutics: they do not even constitute one of its necessary conditions. The therapeutical properties and relations of all the substances and agencies of the materia medica, may be investigated and ascertained, and our knowledge of these properties practically applied, without and wholly independent of any acquaintance with their natural history.

I am not saying that a knowledge of the natural history of the substances and agencies of the materia medica is unnecessary or unimportant: there are various and good reasons why this knowledge is necessary and important; but these reasons are not to be found in any direct connexion between this knowledge and that of their therapeutical properties. Every physician should have some general acquaintance with the natural history of the poppy, and with the physical properties and chemical constitution of opium, but this acquaintance will throw no light upon its action and properties as a remedial agent.[4]

Proposition Second. All science is, in its, nature, positive and simple, in proportion to the positiveness, simplicity and conformity of all the proximate phenomena and relations, which constitute its elements and materials.

I have introduced this general proposition, as the foundation for a few remarks upon the character of the natural in physical sciences, and upon certain differences amongst them; hoping by this means to render the immediate subject of my essay,—the character of therapeutical science—more intelligible and obvious, than it could otherwise be made. By the terms *proximate phenomena* and *relations*, I mean those phenomena and relations, which are nearest to us,—which are cognizable and appreciable by the senses directly, or through such means and appliances as will render them so. Beyond these phenomena,—lying back of them, beyond our reach and beyond our vision,—there are other phenomena, connected to these by relations which we cannot see: and there may be many such of these phenomena and their relations extending backwards from the first until we reach the great, single, primeval force or phenomenon, in which all the subsequent series are contained, from which they all flow and upon which they all depend.

So far as this constitution of the several natural sciences is concerned, there are very wide differences amongst them. In some of them, the proximate phenomena and relations, with which we deal, approach very closely to the ultimate forces, or phenomena, upon which they depend; in others this is not the case,—the primal forces or phenomena, and the proximate are more or less widely separated from each other. The two extreme points in the series of the sciences, arranged according to these differences, are occupied, one of them by the science depending upon gravitation, and the other by the sciences of life. In the first of these, the proximate phenomena may be said to touch the primal force from which they flow. They lie in immediate contact with it, and are its direct manifestation. Nothing intervenes between it and them; at least nothing intervenes to disturb their direct relationship. This primal force of gravity is absolutely independent of all other forces; it is never acted upon or modified, in any manner or degree, by any other forces; it obeys always and absolutely, its own unvarying and unchangeable laws. This great quality of the force itself, and its direct relationship to the phenomena which flow from it, are the conditions which give to the science of gravitation its character of absoluteness and simplicity.

Nothing can be more widely different from all this than the constitution and nature of the science of life. Here the proximate phenomena and relations are removed, more or less remotely, from the primal force or forces from which they flow. The phenomena and their original cause are not necessarily in immediate relations with each other; other phenomena and relations intervene between the primal force, and the proximate phenomena. The primal force itself is not independent and isolated: it is mixed up with other forces: its is influenced and controlled and disturbed by them. The result of these two conditions is a very great variety and complexity in the proximate phenomena of the science. These phenomena are the result of such a variety of forces, working through many successive series of operations, acting and reacting upon each other, that they defy all complete and absolute analysis. The phenomenon before us, the result of all these forces, and so complex in its composition, we cannot resolve into its ultimate elements, or refer to its causative forces. We must deal with it as it is—in the gross.

The primal forces, of which I have spoken, are in their nature, their modes of existence and of actions, entirely unknown to us. We admit their existence and agency because we cannot help it: the recognition of this existence and agency is a necessity of the human intellect: the mind refuses to admit the existence of phenomena that are not dependent upon some force or agency. Still, as I have just said, beyond this, these forces, or agencies are wholly and absolutely unknown to us. This is true alike of them all, if indeed they are many,—of gravitation, of light, of heat, of electricity, of the chemical and the vital forces, and so on. They all lie without and beyond that boundary that separates the known from the unknown.

There is only one other remark that I wish to make in regard to them, when we consider attentively these forces, so incomprehensible and so inscrutable in their nature—so silent in their actions—so stupendous in their power—and issuing, as they do, in such wonderful and manifold forces, of adaptation, order, beauty, and beneficence,—it seems to me to be the dictate of the soundest and most rigorous philosophy, to regard them not as the servants but the masters of matter,—nay more, not merely as blind and inexorable forces, but as direct emanations[5] of the Supreme Intelligence, the living, immediate and very breath and power of God.[6]

Proposition Third. In the Science of Therapeutics, the proximate phenomena and relations, connected with, and dependent upon, one of its two constituents—the articles and agencies of the materia medica—are, virtually and substantially, positive, simple and uniform; and so far the science itself partakes of the same character:—But, the proximate phenomena and relations, connected with and dependent upon the other of its two constituents, the morbid actions, tendencies and conditions of the human economy, are, to a certain extent, and within certain limits, wanting in positiveness, simplicity, and uniformity, and so far the science itself is wanting in the same characteristics. This want of positiveness and absoluteness is not the result of incomplete or partial knowledge; it is essentially inherent in the very elements of the science.

The first member of this proposition does not need any particular illustration. The element in every therapeutical problem, constituted by the substances and agencies of the materia medica is—*quo ad hoc*—a fixed and definite one. It is not fluctuating and variable. The properties and qualities of these substances and agencies are always the same.

But with the other great element in these problems, this is not the case. This element is not fixed, constant, and uniform; it is variable and fluctuating. Disease is never a definite force or quantity, that can be positively and absolutely measured. This force or quantity is never exactly the same in any two cases.

This want of absoluteness and fixity in the pathological constituent of therapeutical science, however, is confined within certain limits: it is not indefinite in degree and extent. Diseases are, for the most part—with certain exceptions, and qualifications—individual entities,—susceptible of being dealt with, as such: they can be characterized, compared with each other, and so on. They are susceptible of definition and arrangement, like any other objects of natural history.[7] Small pox and pneumonia are just as much distinct individual species, as the oak or the lion is. And although of any given species of disease, no two individual cases can be exactly alike, they are still sufficiently alike to be comparable with each other: we can deal with them as we can with the individuals belonging to any species in other departments of natural history. In no species of any of these departments can there be found any two individuals exactly identical. But the differences between the individuals are not indefinite and unlimited: they never overleap their prescribed boundaries. They never depart so far from their type standard, as to lose their characteristic and distinctive attributes.

I will venture to express the opinion, in regard to the greater number of diseases, whose natural history has been made out with a good degree of completeness, when we have arranged them in their natural varieties, depending upon circumstances and conditions whose influence has been ascertained, that the difference between the several cases of these varieties is less than it is generally considered to be. Ten cases of simple pneumonia, for instance, of nearly the same extent, occurring in subjects free from other disease, between the ages of 20 and 25 derived from the same general class of the population, during the same season, will constitute a mass of individual instances which can be very safely and very rigorously dealt with, as elements of scientific research.

I have one other remark to make, and that is there is a great difference amongst diseases, in the degree and constancy of the similarity between their several individual cases. The specific, individual element and character are much more uniform, and strongly marked in some diseases than they are in others. The modifying and disturbing causes are much stronger in some than they are in others. And, finally, there are some morbid conditions, so indefinite, uncertain, and variable in their manifestations as to lose almost entirely this specific and individual character.

I suppose this subject hardly needs any further argument or illustration: I suppose there is no considerable difference of opinion, amongst medical men in relation to it. And yet I am old enough to remember the time when this was far from being the case. One of the chief glories of that famous system, known as physiological medicine, which overran so great a portion of the medical world, twenty-five years ago, consisted in its *disessentializing* diseases ;—in its denying to them the possession of anything essentially specific, or individual; in its reducing them all to one or two, and nearly all of them to a single element. It is always upon this *miserable ontologism* of the nosologists, as he scornfully terms it, that the great founder of this system[8] pours out the hottest vials of his wrath; and he deploys upon it, a power of argument, a splendor of rhetoric, an audacity and brutality of criticism, nowhere else equalled or approached in the annals of medical literature.

The first essential and fundamental condition of all therapeutical science, is to fix as far as possible, this its variable and fluctuating element. The problem to be solved is this: given a certain pathological condition or process, and a certain substance or agent, or a combination of substances and agencies of the materia medica,—to find the true relation between them,— to ascertain the changes effected in the former by the latter. The great, primary, practical difficulty in the way of the solution lies in the want of fixity and absoluteness in one the constituents. We render the solution possible and complete just in proportion to the degree of fixity and absoluteness that we give to this constituent. In other words, the more nearly the several individual cases of disease, with which we are dealing approach each other—the more exactly alike they are—the more nearly they represent equal quantities or forces, the more absolute and complete will our solution become. The means for securing as far as possible, this essential primary condition, may be thus briefly stated.

First: The general hygienic conditions and history of the subjects of the diseases should be substantially the same. They should belong to the same general class of the population; they should have been engaged in the same general class of pursuits; they should have lived in the same general locality. It would not be safe, for instance, to compare any two modes of treatment of typhus, one mode tried upon the comfortable and educated classes—and the other upon the inmates of an alms house. Into what enormous error should we be led, in comparing two methods of treating pneumonia,—one of them in cases of temperate, and the other in cases of intemperate subjects!

Second: The subjects of the disease should be within certain limits, of the same age. Age constitutes a very important element in the natural history of disease. It has especially a very important bearing upon its gravity, and its consequent danger and mortality. Other things being equal, how entirely unlike each other two cases of pneumonia—one occurring between the ages of ten and twenty, and the other between the ages of sixty and seventy.

Third: The subjects of the disease should be of the same sex. This condition is very much less important than the second, but it is not without its value.

Fourth: There may be certain special conditions which are to be considered, for instance, the state of pregnancy. The researches of Mr. Grisolle go to show that this condition adds almost indefinitely to the chances of a fatal termination in pneumonia.[9]

Fifth: The general extent, severity, and character, of the cases, should be substantially the same. It would be very unfair, and could lead to none but the most erroneous conclusions, to compare the treatment of single with that of double pneumonia; of simple with that of malignant scarlatina; of distinct with that of confluent small pox.

Sixth: The period of the disease at which the treatment is commenced, and during which it is continued should be substantially the same. It is quite unnecessary to insist upon or to illustrate the importance of this condition.

Seventh: The method of medication should be as simple as possible and the hygienic circumstances surrounding the patients should be substantially the same.

Finally, after having fulfilled all these conditions, our observation must embrace an adequate number of instances. This adequate number will vary under different circumstances, and in different diseases; but in most cases, as a general rule, it must be large; and the larger the number the more positive and certain will our conclusions become.

I wish to make two other remarks in connexion with this subject—. First, these observations can usually be best made in hospitals; it is, for the most part, in these institutions that the necessary facilities and conditions are to be found. I do not say this absolutely: I admit that it has its exceptions and qualifications: and further that the observations made in these institutions ought to be compared with those made under other circumstances and conditions.

Second, these observations must be made by competent and qualified observers. I need not speak of the necessity of adequate knowledge and skill; but there are two other qualities essential to a competent and trustworthy observer, to which I wish to refer. The first of these is what may be called scientific probity or integrity—truthfulness—supreme love of the truth;—unqualified and unswerving allegiance to the truth—whatever the scientific focus and consequences might be. Our science has always suffered, and suffers now, from the frequent absence on the part of the observer, of this thorough scientific fairness, honesty, and uprightness.

Closely allied to this is what may be called scientific indifference to the results of our observations, a quality of mind of most rare and difficult attainment, but most essential to the trusty and true observer. He must clearly see that the scientific truth he seeks is in nature—not in his thoughts or wishes—and that his sole function is to find out where and what it is. The various passions of the human heart may dread or may desire, this issue or another:—but science asks one only question. And that is—*What is?* And she waits with passive and sublime indifference for the answer, whatever it may be.

Proposition Fourth. The phenomena with which the Science of Therapeutics concerns itself, and which constitutes its material, consists of the changes in the economy, and especially in its diseased states and actions, which are produced by the application and action of the articles and agencies of the materia medica.

This proposition contains a simple statement, requiring no further remark.

Proposition Fifth. Therapeutics is a science of pure observation. Our knowledge of the facts and phenomena which constitute its materials is the result of direct observation of these facts and phenomena themselves; it cannot be derived from any other science: it is not the result of any process of pure logic, or of à priori *reasoning:—it is not contained in, and it cannot be inferred or deduced from any other phenomena than its own; it cannot be deduced form pathology; it is not a corollary*[10] *of pathology; it is not contained in, and it cannot be deduced from the physiological relations of the articles and agencies of the materia medica, or their actions upon the healthy system:—it cannot be deduced from comparative therapeutics, or the action of these articles and agencies upon the diseases of the lower animals.*

This Proposition contains several very important doctrines.

Therapeutics is a science of pure observation. I hope there is no need, at this time, that I should argue this proposition. I wish merely to insist upon it in its absoluteness, comprehensiveness and completeness. I wish to insist that it shall not be qualified, or evaded, in any of its parts, or in any of its consequences. Therapeutics is not the result of any speculations or merely logical inductions:—it is not the issue and product of any process of *à priori* reasoning.

Therapeutics is not contained in pathology. It does not flow from pathology, independent of and antecedent to, direct observation. A knowledge of pathology does not, in itself, comprehend and include a knowledge of therapeutics.

Let me not be misunderstood. There is a relation—intimate and indissoluble—between pathology and therapeutics: there is a certain sense in which therapeutics may be said to rest upon pathology. This has already been stated. One of the two constituent elements of all therapeutical science is the natural history of disease,—embracing its pathology. The first absolute and indispensable condition of all therapeutical science is the diagnosis of disease; and an indispensable element of diagnosis is pathology.

The science of therapeutics does not rest upon, and it cannot be derived from the actions of the substances and agencies of the materia medica upon the healthy system. What is the science of therapeutics? What does it deal with? And what are its phenomena? It is the science of *curative* relations: it deals with pathological and not with physiological processes and conditions.

I do not deny the utility of a knowledge of the physiological relations of the articles of the materia medica—of their action upon the healthy system. This knowledge is very important: it frequently sheds light upon their curative properties:—I only insist that it does not constitute a legitimate element of the science of therapeutics; and that its indications are to be received with caution, subject always to the authority of direct therapeutical observation.

Therapeutics is the science of curative relation, in the *human* economy: it is not contained in the curative relations of the lower animals. Therapeutics is one science: Comparative therapeutics is another. I do not deny the utility of the latter; let the actions of the articles of the materia medica —upon the healthy and the diseased conditions of the lower animals be unmitigated[11], and ascertained:—this knowledge may throw some light upon their true physiological and therapeutical relations: but it does not constitute a legitimate element of human therapeutics: its conclusions should be received with great caution,—much greater than that with which they generally have been received; and subject always to the revision and adjudication of the only legitimate and supreme authority—that of direct therapeutical observation, made upon the human economy itself.

It is true that a knowledge of the chemical properties of certain substances, of their physiological action, and of their effects upon the diseases of the lower animals, may lead to certain hypotheses, more or less probable, or what may be their therapeutical properties. But what I insist upon is that these suppositions are not science. It is only by direct observation of the phenomena themselves, of the action of these substances upon human diseases, that their therapeutical action can be positively known; that they can become a portion of therapeutical science.[12]

Proposition Sixth. In order that the phenomena of therapeutics may be erected into a body of science, their true relations must be ascertained; this can be done only by careful and rigorous analysis and classification; and in order that these processes may be adequately performed, there must first be a true record *of the phenomena. It is not only necessary that these phenomena should be, as far as possible, weighed and appreciated, but it is also necessary that they should be enumerated, as far as they are numerable.*

The foregoing conditions having been, all of them—fulfilled: every precaution having been taken to arrive at the truth: every known source of error having been as far as possible eliminated, how are we to ascertain and sum up the result of our labor? Will these results issue spontaneously from this labor: or can they be evoked from it by any simple act of the judgment or the will? Is there any degree of sagacity or good sense, or sound judgment, that is adequate to do this? Most certainly not. Unaided by any other processes, and trusting to its general memory of what had been done, the mind might indeed form some general notions of these results; it might venture issue some conclusion of greater or less probability: but it can do no more.

In the first place, before we can do anything with these observations, we must have them before us. They have been collected at different times, in various places, under various circumstances; we must now gather them together, and we must have them in their completeness and integrity. In order that we may do this, these phenomena must have been recorded,—fairly and fully recorded at the time of their occurrence. It should be an axiom in the philosophy of therapeutics, that all facts and phenomena, not recorded at the period of their occurrence, and trusted entirely to the unaided memory, are almost certainly without value, as elements of positive science. If these facts and phenomena are few in number, they are, for this reason, of little or no value; if they are numerous, it is impossible for the memory to retain them.

Having thus got our facts and phenomena fully and fairly before us, the great process to be brought upon them is to ascertain their true relations. In order to do[13] this, they must be analyzed, arranged, and classified; and these processes must be continued and repeated, carefully, conscientiously, and rigorously until[14] these relations are ascertained, until the accidental, the fortuitous, the apparent merely are sifted out and separated from the natural, the actual and the true. Now, whatever other conditions may be essential to these processes, this, at any rate is essential, that we should know *how many* the facts and phenomena are, with which we are dealing. *They must be counted.* All the facts and phenomena that are numerable must be enumerated. Suppose we are studying the therapeutics of pneumonia. Certainly before proceeding to the analysis of our cases, we must know how many there are:—we must count them. And so of all the elements of these cases. We must know how they are divided between the two sexes:—and to know this we must count them again. We must know how they are distributed among the several periods of life; and we can know this only by counting. We must know how many terminate in death and how many in recovery: in order to know this we must count them. We must know how often certain pneumonia occurred; how frequently a certain event followed some other certain event, and all this can be done only by counting. All this is to my mind so perfectly obvious, that I hardly know how to go to work to enforce it by any formal argumentation, or illustration. Still, it involves the whole doctrine of the numerical method in its application to therapeutics; and this application has been and still is so strongly opposed, that I wish to make one or two remarks in relation to it.

It seems to me that the objections which have been made to the use of this means,—to the application of this analytical process to therapeutical phenomena, has grown out of a misapprehension of the nature, purposes, and functions of the process. If this is not so, then the objections are to me unintelligible. It ought to be distinctly seen that this method has nothing whatever to do with the observation of therapeutical phenomena: it has no connexion with this observation; its functions do not commence until the work of observation has been already completed. It deals only with the dead results, if I may so say, of observation; it simply takes up the gross, aggregate mass of phenomena, after they have been ascertained by observation and recorded, and this for the sole purpose of interpreting their meaning, of ascertaining their real significance, of getting at their true value. It has indeed one direct relation to the observation and study of therapeutical phenomena, and that is this:—it requires that the rigorous conditions that have been already stated should be vigorously fulfilled:—it requires that the observations with which it is to deal should be carefully, thoroughly, conscientiously, adequately made: it can deal with no others: it is only the rough ore of such observation, that is either worthy or susceptible of transmutation in its crucible into the pure gold of science. And it is in *its* crucible only, that this transmutation can be wrought: in the old alembic,—with the old reagents,—by the old processes,—with scanty and inadequate materials,—trusting to the general memory—relying upon sound judgment, or good sense, or great genius alone—a very plausible and promising result may easily be obtained,—but this result will be, at least it may be, spelter and not gold.

This reflex action of the process of analysis is a very noticeable and striking fact; and very important, also, in its influences and results. By rendering necessary to its own ends and purposes, the fullest and completest possible observation, it secures and guarantees this observation. Its own functions are powerless, unless the rigorous conditions of their actions are first fulfilled: it can do nothing itself until its own exacting demands have been complied with. The loose and inadequate data that serve for the conclusions of the unaided judgment and common sense of the observer are to it of little or no value.[15]

One of the greatest objections that has been made to the application of this method to pathological and therapeutical phenomena has grown out of the fact of its attempting, sometimes, to deal with these loose and inadequate data. Whenever it attempts to do this; or whenever it fails or refuses to deal adequately and fairly with its adequate materials, its conclusions are sure to be erroneous, its decisions are necessarily fallacious and false.

The remarks contained in the two last paragraphs are also true of the numerical method of analysis and appreciation in all its various and manifold applications,—to the science of life in every department,—to social and political economy,—to practical astronomy,—to meteorology, and so on. All these sciences are such, on two absolute conditions;—first, an adequate observation of their phenomena, and second, the ascertainment of the real value of these phenomena, of their true significance and relations, by means of numerical analysis and estimate. Either of these conditions failing, or being defective, the science itself fails, or is defective in the same degree; and the science becomes complete, determinate, and positive, just in proportion to the completeness with which both of these conditions are fulfilled. It is true also of all these sciences, as it is of therapeutics, that the numerical method, in its application to them, in order to the performance of its own functions, and the accomplishment of its own ends, must be furnished by observation with adequate and authentic materials; it cannot deal legitimately with any others. Statistical science, in all its applications, fails of its end, frustrates its own purposes, and becomes false and fallacious, only when the materials upon which it works are spurious or defective, or its own work is inadequately or dishonestly done.

Let us take any one of the great, important, ever-recurring problems in therapeutics, and see if there is any other method by which it can be efficiently dealt with, and satisfactorily solved. Acute inflammation of the lungs is a common disease. Its natural history,—its pathology—its signs and symptoms—its duration—its tendencies—its degree of danger—the circumstances which form its occurrence, have all been very well ascertained. Now beside all this, we wish to know what its therapeutical relations are. We wish to know, for instance, what effects are produced upon it, what amount and degree of control over it are possessed, by bloodletting: not whether *any* effects are produced upon it, but as accurately and as positively as possible, *the nature and amount* of these effects. How can this be done? There are only two general methods, and which of the two is adequate to the performance of this work? Is the old method adequate? Will the life-long observations of a Huxham, a Sydenham, a Nathan Smith,[16] a Latham,[17] trusted to the unassisted memory, unrecorded, unarranged, unanalyzed, uncounted, ever definitively and peremptorily settle this question? Carried on, as they have been—by multitudes without number of observers—from the age of Hippocrates to that of Louis—did they settle and determine this question? did they solve this capital problem? did they give to it that rank and character of scientific certainty of which it is susceptible? You all know there is but one answer to these questions; differing only in the degree of emphasis, that will be given to it, by all competent judges.

The other method was first adopted and fully carried out more than twenty years ago by Louis—some of you may remember the general chorus of incredulity, distrust, and in some instances of fierce[18] and scornful hostility, with which the simple, modest, manly and truthful statement of the results of his investigations was received by his contemporaries. Amongst other things it was said, that he rejected bloodletting in pneumonia, and denied its utility; and now, when there is danger that opinions may run to the other extreme, to what recognized authority do we appeal but to him, and to that of his worthy disciples who have followed the same methods? So the great central luminary of our system calls back the planets from their widening excentricities and holds them by its silent, but mighty power within the orbits that were ordained for them from the beginning. The observations of Louis, of James Jackson, of Grisolle, and of others, made in the same spirit, subject to the same conditions, have—I do not say absolutely and definitively solved this problem and settled this question—but they have furnished the first legitimate and adequate materials for their solution and settlement: they have adopted the only method by which this solution and settlement can be achieved.

How is it with many others, of what may be called leading and capital questions in therapeutics? Are they settled to your satisfaction? Are you content with the present state of our knowledge in regard to them? How is it in regard to the effects of our great, heroic remedies, upon the principal forms of grave, acute disease? Has the Hippocratic method of observation, pursued for two thousand years, been sufficient to determine these matters? Are the effects of blood-letting, opium, and calomel, upon the several simple acute inflammations, satisfactorily ascertained, and determined,—so as to be matters not of varying, and contradictory opinion merely, but of settled and determinate science? What is the influence of opium over acute peritonitis, and pericarditis and meningitis? What is the best treatment of membranous croup? And so on.

It seems to me hardly necessary that I should pursue the subject any further. I wish, however, to make a single qualifying remark. I do not pretend that this rigorous method is applicable to all therapeutical phenomena. There are many morbid elements and conditions not amenable to the laws of this method. They are too fluctuating and variable; they have in them too little constancy, fixity, and individuality, to be thus dealt with, and we must be content to deal with them in a looser[19] and more general manner.

Proposition Seventh. The chief inherent reason of the imperfection of the science of therapeutics, and the great obstacle to its advancement, consists in the difficulty of distinguishing between those changes in the condition and progress of disease, which are spontaneous, or independent of the actions of remedial means, and those which are the result of this action. Or, in other words, the reason of this imperfection, and the obstacle to this advancement, consists in the difficulty of ascertaining the existence and the true relations of therapeutical phenomena.

Disease is not often a stationary process or condition. Almost always, it is more or less progressive, in one direction or another; it is subject to changes and complications—more or less regular and constant—more or less sudden—more or less important. The quality of disease renders it extremely difficult, in many cases, to distinguish between those changes which are effected by remedies, and those which are independent of them. This difficulty is especially great in that large class of diseases which have a tendency, after having passed through a certain series of processes, to terminate in recovery. There is only one means of overcoming this difficulty, and that consists in a thorough knowledge of the natural history of disease. We must know what the changes are, which more or less constantly, under certain circumstances, and at certain periods, occur in disease, independent of remedies, before we can distinguish between these, and those that are really therapeutical.

It is not necessary to enlarge upon this subject. There is no greater difficulty than this in the way of scientific and practical therapeutics: there is no more constant or frequent source of fallacy and error. For how many years—throughout the medical world—was the sleep which usually follows the prolonged wakefulness of delirium tremens, confidently attributed to heroic doses of opium!

Proposition Eighth. Since one of the two constituent elements of the science of therapeutics is, within certain limits, indefinite, variable, and fluctuating, the science itself must partake of the same character: it is only approximative; it is not and cannot be positive and absolute.

The science and the art of therapeutics are approximative and not absolute. The philosophical grounds and reasons of this have been already sufficiently stated. This character of the science and art is inherent: it is not the result of imperfect knowledge.[20] The means of diminishing as far as possible this approximative and contingent character, and of giving to the science and the art the highest degree of positiveness and certainty, of which in their nature and constitution, they are susceptible, are to be found in the methods and conditions which have been already stated.

Proposition Ninth. A law or principle of therapeutics consists simply in a generalization of some of the facts, phenomena, events, or relations, by the sum of which the science is constituted. Like one of their classes of constituent elements, they are within certain limits, variable, and fluctuating: they are approximative and not absolute. The degree in which these laws or principles approach to absoluteness and positiveness, varies very greatly; this degree is in proportion to the constancy and invariableness of the particular class of phenomena or relationships constituting the elements of the law, and to the extent and accuracy with which they have been ascertained.

I have no time for the full development of the several members of this proposition. The nature and constitution of a law, or principle, in therapeutics is sufficiently fully and clearly stated. These laws belong to the class of generalizations that are called empirical. Like the elements by which they are constituted; like the phenomena which are their materials, they are approximative and not absolute.

In therapeutics, as in all the other departments of the science of life, we cannot go one step beyond these first generalizations—these empirical laws. The vital forces, upon the actions and relations of which, the phenomena constituting the materials of these laws depend, are beyond the reach of human knowledge; at any rate they have thus far eluded all our researches and investigations. We cannot do here, as was done in astronomy, advance from the lower empirical laws to the simple and ultimate forces from the action of which these laws have flowed, to that great, absolute higher law, which contains and includes all laws below it, from which all these issue and upon which they depend. The work to be done in the science of life, and in therapeutics, as one of its branches, is the work of Kepler, and not that of Newton.

My faith in the powers and capacities of the human understanding is as high and stirring as that of any one; it is as puerile as it is unphilosophical, to attempt to limit or define the extent of these powers: but certainly they have limits. The human understanding is not omniscient: the luminary or shadowy circle of the known, within which we stand, rests now and will rest forever upon a background of the unknown, wholly dark and impenetrable. It seems to me that all our present knowledge of life and the soundest and soberest philosophy leads to the conclusion that the vital forces, and their primary actions are beyond this line. The chemist has no alchemy so subtle, the microscopist has no vision so specific and penetrating, as to detect, even so shadowy or so crudely the differences between two ova that are before him: and yet while one of them contains within its minute circumference, potentially a Shakespeare or a Newton, the other holds a jelly-fish.[21]

The means of rendering these laws and principles as nearly fixed and positive as their nature and constitution will admit are to be found in the methods and processes already laid down,—applied to a sufficiently large number of instances. This indispensable condition—in addition to all the others—of large numbers of instances—of very considerable, or great aggregate masses of material—and the manner in which they are to be treated has been fully stated and developed by Gavaret.

Nous ne savons le tout de rien—We do not know the whole of anything—said Montaigne.[22]

Proposition Tenth. There is only one true, philosophical indication, in therapeutics; and that is the removal or mitigation of disease: when any other signification is given to the term, it becomes either hypothetical in its meaning, or inadequate and erroneous in its application.[23]

It has always been a common and favorite doctrine in medical philosophy, that therapeutics ought to be founded upon, and to be guided by, what are called rational indications. There seems to have been a sort of vague and indefinite notion, that, in this way, the art could be rendered more rational, more scientific, more philosophical, and less empirical, than it would otherwise be. Thus, for instance, John Brown would have said, that the great, leading, rational indication, in the treatment of nearly all diseases, consisted in the removal of their asthenic element. The old humoral pathologists placed their indications in the coction, elaboration, and removal from the blood of its peccant humors. I wish merely to say that the intervention of this element, in the form of a rational indication, between the remedy and its proximate, appreciable effects is wholly gratuitous, and almost always hypothetical.

Proposition Eleventh.[24] *The* modus operandi*, as it is technically termed, of remedies, is, for the most part, unknown to us: the term when applied, as it usually is, to the primary action of remedies, is for the most part, wholly hypothetical.*

With certain not very numerous and not very important exceptions, the *modus operandi*, as it is termed, of medicines is not a subject of positive knowledge. We know very little of the *how* of remedial processes; we know very little of the primary, and essential actions of most of the articles and agencies of the materia medica. I am not speaking of course of some of the grosser and more mechanical of these processes—the removal, for instance, of irritating substances form the alimentary canal by the aid of cathartics or emetics—but as a general rule it is true, that the appreciable effects of remedies are the results of actions and processes whose intimate and essential nature is wholly unknown to us. *How* does antimony diminish the intensity of inflammation of the lungs? By what primary and essential action does cinchona modify and arrest the processes of periodical disease? By what mysterious modification, indelibly impressed upon the economy—and upon what element of the economy—the solids—or the fluids—one or many or all—does the vaccine disease destroy the susceptibility to small pox? For what special and peculiar agency does opium cure acute inflammation of the peritoneum?

Closely connected with this subject, there is another matter, in relation to which it seems to me important that we should have clear and correct conceptions. It is a popular doctrine of the current medical philosophy, to deride and deny the special action of remedies; the very term specific has come to be, as you well know, a term of reproach. We are told that the curative effects of remedies in individual diseases, must be referred to certain general actions and properties, and not to any special or peculiar relation existing between the remedy and the individual disease. An attempt has been made to make general therapeutics, the source and basis of special therapeutics. This has arisen from the same tendency which I have already had occasion to combat—the imposition and effort to render therapeutics *rational*, as it is termed, and to free it from the reproach of empiricism.

Now it seems to me that a sound philosophy, and a thorough knowledge of diseases and their remedies furnishes no countenance or sanction to this doctrine. On the contrary, it seems to me that this philosophy and this knowledge render very manifest, and very prominent, the special character of individual diseases, and the special action of individual remedies. I do not deny that many diseases have certain common elements:—I do not deny that many remedies have certain general activity:—I only insist that in nearly all—perhaps in all—diseases—even in those where there may be most of these common elements—there are also special and peculiar elements, which give to them a special character: neither do I deny that many remedies have certain general actions; I only insist that these remedies have also special and peculiar properties, and special relations to certain individual diseases. It does not necessarily[25] follow, that that because bloodletting has been found to have a curative influence over simple acute inflammation of one organ or tissue, it must necessarily have the same influence over others. It does not follow that because antimony controls pneumonia, it will also control hepatitis. It does not follow that because opium cures acute inflammation of the peritoneum, it will also cure acute inflammation of the pia-mater, or the pericardium. I think it is a dangerous as well as a false doctrine, which teaches us to disregard these special elements in disease, and in their remedies, and to underrate their importance.[26]

Proposition Twelfth.[27] *The end and object of the Science of Therapeutics may be stated in the following formula:—Given a disease, or a morbid condition of the human economy, to find a remedy, or remedial method.*

This proposition needs no development. The science of medicine issues finally in the end of therapeutics. This is its consummation.—its great end and purpose. Anatomy, physiology, pathology,—the entire natural history of disease—materia medica—all are preliminary, more or less direct and essential, for the cure or the mitigation of disease. It is true that medicine may be studied, simply as a branch of pure natural science, without any reference to its application as an art. But this is not often done. The great purpose of this study is to make the physician; and the physician is he who, within the limits and conditions of his science and art, prevents, mitigates and cures disease.

NOTES

An Essay on the Philosophy of Medical Science

Preface

[1] Pierre-Charles-Alexandre Louis (1787-1872) spent six years studying disease in the wards and autopsy rooms at La Charité in Paris. His method of meticulous recording of his observations and his use of the numerical method were central parts of the Paris clinical school. Louis was the major influence on the Americans who went to Paris to study medicine in the early nineteenth century.

[2] Jean-Baptiste Bouillaud (1796-1881) was a French physician and researcher. One of his major accomplishments was localizing the speech center in the brain. Bartlett here refers to Bouillaud's 1836 *Essai sur la philosophie médicale et sur les généralités de la clinique médicale, précédé d'un résumé philosophique des principaux progrès de la médecine.*

[3] Sir Gilbert Blane (1749-1834) was a Scottish physician. Blane had sailed to the West Indies in 1779. Through his influence, in 1795 the use of lime juice became required throughout the navy to prevent scurvy. Among his works are *Observations on the Diseases of Seamen* (1795) and *Elements of Medical Logic* (1819).

[4] William Hillary (1697-1763), an English physician, is best known for his studies of the effects of weather on diseases in Barbados.

[5] Théophile Charles Emmanuel Édouard Auber (1804-1873) also published his *Philosophie de la médecine* in 1865.

[6] Vincenzo Lanza's (1784-1860) book appeared in an English translation from the Italian by C. Stormont in 1826.

Part First

Chapter I

[1] Bartlett seems to be making a psychological claim here: the facts of physical science will be more obvious to readers even though there are no essential

ontological or epistemological differences between physical science and medical science.

[2] Bartlett is here decrying the imprecise use of the term “inductive reasoning.” It is unclear, however, whether he is condemning inductive reasoning, understood as reasoning from specifics to a generality. Bartlett never clearly describes the reasoning process he would allow in arranging and classifying facts and phenomena.

[3] This is Bartlett’s ontological claim about the nature of science. Science just is a collection of facts and phenomena. There is no room for explanation in this view of science.

Chapter II

[1] Original edition has “aranged.”

[2] While it might be argued that gravitation is a constructed theory that explains why objects fall, Bartlett takes gravitation simply as the sum of the observed facts of falling objects.

[3] Sir Isaac Newton (1642-1747) is well known for his work in mathematics and physics.

[4] Sir William Herschel (1738-1822), discoverer of Uranus, detected heating rays beyond the visible red of the spectrum in 1800. After his death, his son, Sir John Herschel (1792-1871), made the first thermal image from sunlight.

[5] Étienne-Louis Malus (1775-1812), was a French artillery officer and engineer. He discovered the polarization of light by reflection from a surface.

[6] Augustin Jean Fresnel (1788-1827) was a French physicist. He supported the wave theory of light and developed the compound lenses used in lighthouses.

[7] Thomas Young (1773-1829) was an English physicist whose wave theory explained the interference that is observed with light.

[8] Joseph-Louis Lagrange (1736-1813) is usually considered to be a French mathematician although he was born in Italy. He developed many areas of pure mathematics.

[9] Pierre Laplace (1749-1827) was a French mathematician and physicist who worked in mathematical astronomy.

[10] William Wollaston (1766-1828) was an English chemist and physicist.

[11] Sir David Brewster (1781-1868) was a Scottish physicist.

[12] This is an example of what we would today call hypothesis testing, and is clearly acceptable to Bartlett. For Bartlett, however, the science is not in the process of testing, but just is the collection of observations that result.

Chapter III

[1] Bartlett here seems to depart from the empiricism espoused by British philosophers such as David Hume by making a metaphysical assumption about the regularity of the workings of nature. One can read Bartlett as being a naïve realist and simply assuming that our senses tell us the way the world is in reality. However, Bartlett's footnote about cause and effect indicates that he is simply leaving aside the question of whether the observed regularity might be a result of the constitution of the human mind. Hence, we can plausibly conclude that Bartlett is aware of the deep philosophical issues raised by his epistemological stance, but he thinks that he can make his point without solving the deeper metaphysical problem.

[2] The metaphysical question about the nature of a scientific law is raised here. Bartlett equates laws with generalizations about observed phenomena. Laws are not separate metaphysical entities standing behind the phenomena and somehow causing them. Laws are simply universal facts themselves.

Chapter IV

[1] That Bartlett never precisely defines "hypothesis" was an obvious objection for the first reviewers of the *Essay*. Roughly, he uses "hypothesis" to mean an attempted scientific explanation not based upon observation. What is clear, however, is that Bartlett does not employ the term in the way that present scientists do—a conjecture to be submitted to empirical testing for ultimate acceptance as fact or rejection as falsity.

[2] Physical science now recognizes several types of subatomic particles, and offers empirical evidence for their existence. Bartlett's use of the term "atom" in this particular context ought to be construed in a philosophical sense; that is, as the smallest individual unit of matter. When he later talks about chemistry, we can safely take him to be using "atom" in the more customary sense.

[3] William Whewell (1794-1866) was an English polymath. The first edition of his *History of the Inductive Sciences* was published in London in 1837. Bartlett quotes extensively from this work. Whewell claimed to be a follower of Lord Bacon and to be "renovating" Bacon's method of induction. Whewell criticizes both the idealism of Kant and the empiricism of Locke and tries to find a middle way to scientific knowledge. Whewell held that "fundamental ideas" such as space, time, cause and resemblance are not gained through experience, but are supplied by the mind and

enable the organization of the data of our sense experience. It is unclear how much of this fundamental epistemological position Bartlett accepts. Bartlett never delves into such basic matters. If he presumes such a foundation, however, his brand of empiricism begins to look much more sophisticated than Bartlett's critics give him credit for.

[4] Bartlett's suggestion here is not far removed from the ideas of contemporary philosophers of science who espouse Thomas Kuhn's idea of paradigm shifts in scientific revolutions. Contemporary science does include many of the kinds of explanatory structures that Bartlett would remove from it.

[5] Joseph von Frauenhofer (1787-1826), a German physicist, investigated the spectrum of the sun.

[6] Christiaan Huygens (1629-1695), a Dutch physicist, was a leading proponent of the wave theory of light.

[7] Ottaviano Fabrizio Mossotti (1791-1863), an Italian physicist and mathematician, worked on the foundations of molecular physics.

[8] Sir Humphry Davy (1778-1829) was an English chemist who experimented with gases by inhaling them. This sometimes made him quite sick, but also led him to discover the intoxicating anesthetic properties of inhaled nitrous oxide.

Chapter V

[1] Science, for Bartlett, is constructed by classifying and arranging observed facts and phenomena. The principles given for the classification and arrangement of facts, however, are less clear than Bartlett supposes. The degree of similarities and dissimilarities between phenomena is at best a non-specific criterion for the construction of a field of study so complex as science. Bartlett does not tell the reader what is to count as a similarity. Specifying further the principles necessary for the act of classifying and arranging would seem to require value judgments about relevance of criteria, degrees of similarity, etc., and these sorts of judgments are intrinsically outside the realm of empirical study.

Part Second

Chapter I

[1] Bartlett is careful here to note that the facts of pathology are ascertained through observation, and not through inference from the facts of anatomy and physiology.

[2] Bartlett here makes therapeutics merely a division of pathology. In his later *Philosophy of Therapeutics*, he severs this relationship. In his discussion of the fifth

proposition of that work, he admits that there is an important relationship between pathology and therapeutics, but specifically states that therapeutics is not contained in pathology and does not flow from pathology.

Chapter II

[1] Just as he claimed for physical science, Bartlett sees medical science as consisting only of a collection of facts. He disregards the role of reasoning in arranging and classifying these facts as well as the ontological status of facts themselves.

Chapter III

[1] Bartlett now condemns deduction in medical science, as he has previously condemned induction. Here, he uses deduction in its proper sense of reasoning from a general principle to a specific conclusion. His use of these terms is not always so precise, however. See pages 212-216 of the original text for Bartlett's further discussion of the use of deductive reasoning.

Chapter IV

[1] In all the examples in this chapter, Bartlett is not making any metaphysical claims about the relation between, for example, gross and microscopic appearance of organs. It may, in fact, be the case that deposits of a certain chemical composition do appear similar even in different organs. Bartlett's claim is merely epistemological. We cannot make any conclusions about chemical composition from gross appearance. Chemical testing is necessary.

Chapter V

[1] For Bartlett, function is ascertained through observation alone. In the preceding examples, Bartlett seems to find function to be self-evident once one simply has observed a particular structure. He makes it clear in his footnote that he is not arguing from apparent function to the existence of an intelligent designer. This may blind him to the fact that discovering a function in a particular structure requires a sometimes sophisticated reasoning, usually by analogy with some other structure that has a similar function. Even when such analogies are straightforward, however, it may still be the case that other less overt functions are also attached to a particular structure. While observation may be necessary to establish such connections, it is not sufficient. Bartlett seems to appreciate some of this complexity only in his remarks about the limitations of observation to establish function. Here, however, his emphasis is on the actual limitations of observation and not on the conceptual limitations of observation alone. He remains blind to the conceptual limitations.

[2] Original edition has "knowlege."

[3] This example points out the complexity of the point Bartlett is arguing, but that Bartlett seems to miss. We would probably not say, today, that digestion occurs in the stomach, although the stomach plays a role in digestion. If by "digestion" one means absorption, digestion occurs in the small intestine, but different parts of digestion also occur in the liver. Such facts were established in a significant way through observations that were impossible in Bartlett's day.

[4] Again, mere observation, although necessary, is not sufficient to establish such facts.

Chapter VI

[1] Jean-Jacques Rousseau (1712-1778), *Les Confessions*, Book 5.

[2] Here Bartlett admits, as a qualification, that one form of reasoning is to be allowed in science: drawing conclusions from well-established laws. The basis for this admission is the metaphysical supposition that some of the workings of nature are perfectly regular. What is not made clear is the epistemological standard by which we can judge that we have sufficient knowledge to proclaim such a law.

[3] Of course, if we know, for example, that a certain area of the brain is necessary for speech, we may reasonably conclude that destruction of that area will result in loss of speech. This does not seem to be Bartlett's point. Rather, it is that simply knowing the functions of the brain in general, without reference to what specific areas are responsible for specific functions, will not enable us to predict, *a priori*, the effects of different sorts of injuries. Knowledge of cause and effect comes only from empirical observation.

[4] These "qualifications and exceptions" admit into Bartlett's philosophy of medical science the kind of reasoning that is obviously necessary if his scheme is to be at all plausible. Thus, it is not all reasoning that Bartlett wants to banish from medical science, but primarily the speculation of the rational systems of his day.

[5] That Bartlett sees these "qualifications" as little more than truisms and merely exceptions to his rule against reasoning in science shows once more that he has not precisely delineated particular forms of reasoning.

[6] Bartlett might have seen these examples and the examples that follow as more akin to the "exceptions and qualifications" he has discussed if he had known more about the structures and functions he discusses. For example, if Bartlett had our present knowledge of the structure and function of the immune system in the various organs, he might have been able to predict the various phenomena of inflammation in the way that he predicted circulatory failure from the destruction of the aortic valve.

None of this, however, is inconsistent with his philosophical stance on empiricism; neither does it lend any support to the sort of speculation Bartlett criticizes.

[7] A type-setting error in the original edition has page 97 numbered as page 79.

[8] Bartlett here makes an interesting and unargued claim about the priority of the normal over the pathological. That is, one cannot recognize a pathological state without first knowing the healthy physiological state. This may have seemed self-evident to him. However, this is a matter separate from the question of how one comes to know pathological states. The latter is Bartlett's concern in this chapter.

Chapter VII

[1] The question of causation is not addressed by Bartlett in any philosophically sophisticated manner. Bartlett simply argues that our knowledge of cause and effect comes from direct observation. This position stands in opposition even to the empiricism of David Hume, who held that although we can observe events, we cannot observe causation. For Hume, causation is in the human mind, and not in nature. We attribute cause and effect by association when we observe that certain events are constantly conjoined in a particular temporal relationship.

[2] Here Bartlett does not distinguish between etiology and epidemiology. Even though certain diseases, influenza, for example, occur more frequently in the winter, season of the year is not an etiological agent, in the strict sense, but merely an epidemiological factor.

Chapter VIII

[1] Bartlett now gives epistemological criteria for using reason to judge when it is safe to conclude that removing a cause of a disease will remedy the disease. The cause-effect relation must be direct and simple, that is, with no intervening phenomena. Knowing that the criteria are fulfilled in a particular situation, however, may be difficult.

[2] Malaria

[3] Scrofula is tuberculosis affecting the lymph nodes of the neck. This may have been confused with goiter, an enlargement of the thyroid due to iodine deficiency.

Chapter IX

[1] Bartlett understands diagnosis to be an art dependent on the science of pathology. What follows concerning the relationship of diagnosis and therapeutics is really more about the relation between knowledge of pathology and therapeutics.

[2] What circumstances are relevant is an important unanswered question. Bartlett's claim of regularity in outcome of giving a particular therapeutic agent is another instance of his unargued assertion of the regularity of the workings of nature.

[3] Original edition has "to day."

[4] This is Bartlett's epistemological constraint on his absolutist metaphysical claim about the regularity of nature. Even though Bartlett assumes that there is a constant relation between particular diseases and particular therapeutic agents, he recognizes our practical inability to say for certain that we are observing a regularity in nature in any given case. Our uncertainty arises not from any irregularities in the relation between disease and therapeutic agent, but only from the unknowable complexity of diseases. In the eleventh chapter, Bartlett will make a distinction between absolute laws and ascertainable laws.

[5] Here is an even stronger epistemological constraint. When a particular disease affects an individual, the unique constitution of that individual modifies the disease itself.

[6] Original edition has "mere."

Chapter X

[1] See chapter XIV.

[2] Bartlett does not give criteria for judging whether particular phenomena should be considered "fundamental and most important." Thus, he sees cancer in different organs as a common manifestation of one disease, but inflammation in different organs as different diseases.

[3] Although Bartlett includes cause as one of the criteria for distinguishing these diseases, he is really making distinctions on the basis of the phenomenal manifestations of the various diseases. The bacterial or viral etiology of the diseases he discusses would not be discovered for some years. The causes to which Bartlett refers are those discussed in the fifth chapter of the second part.

[4] Again, Bartlett asserts a metaphysical certainty but an epistemological uncertainty.

[5] John Armstrong (1784-1829) was an English physician and lecturer at several private medical schools.

[6] Francis Boott (1792-1863) was born in Boston, graduated from Harvard, studied medicine in Britain, and received his M.D. degree from Edinburgh in 1824. Boott also was a noted botanist and wrote a memoir of John Armstrong.

[7] John Huxham (1692-1768), an English physician and poet, published *An Essay on Fevers* in 1750.

[8] Erasmus Darwin (1731-1802), an English physician, botanist and poet, founded the Philosophical Society of Derby in 1783. He was the grandfather of Charles Darwin.

[9] Sir John Pringle (1707-1782) was a British physician who also acted as professor of moral philosophy at Edinburgh from 1734 until 1742. His *Observations on the Nature and Cure of Jayl-Fevers* was published in 1750; his most famous work, *Observations on the Diseases of the Army, in Camp and Garrison* was published in 1752.

[10] Bartlett is here referring mainly to epidemiological factors.

Chapter XI

[1] The distinction between absolute and ascertainable laws depends not so much on a distinction between laws of physical science and laws of life sciences *per se* as on the level of epistemological certainty that the complexity of the laws allows. Bartlett, perhaps naively, takes the laws of physical science to be absolute, relying on something like his "direct and simple" criterion for establishing a relation between cause and effect. Laws in the life sciences, on the other hand, are too complex to allow a judgment that they are absolute.

[2] This is an important point in Bartlett's concept of disease. It is impossible to give conditions that are both necessary and sufficient for a disease. This does not make nosology impossible, however. We can still confidently rely on what Wittgenstein would come to call "family resemblance."

[3] Louis-Dominique-Jules Gavarret (1809-1890) advocated and developed Louis' *méthode numérique*. His *Principes Généraux de Statistique Médicale* was published in 1840.

[4] Bartlett is, in essence, talking about introducing controls into research studies.

[5] Chapters VIII and X.

[6] Pulmonary tuberculosis

[7] Stroke

[8] Georges Cuvier (1769-1832). *Discours sur les révolutions du globe*, 3d ed., Paris, 1825. Introduction.

[9] Hippocrates of Cos (460-377 B. C.) rejected the superstition of primitive medicine and advocated clinical observation to discover the natural origins of disease.

[10] Albrecht von Haller (1708-1777) was a Swiss physiologist who conducted studies on muscle and nerve.

[11] Giovanni Battista Morgagni (1682-1771) was an Italian physician, anatomist, and pioneer in pathology. In 1761, he published *De sedibus et causis morborum per anatomen indagatis*, which established the modern field of pathological anatomy.

[12] Thomas Sydenham (1624-1689) has been called the "English Hippocrates." He advocated careful observation and clinical experience as the basis of medical knowledge.

[13] John Hunter (1728-1793) was a Scottish anatomist and surgeon. He rejected speculation and advocated direct observation and experimentation.

[14] René-Théophile-Hyacinthe Laennec (1781-1826) was the great clinician who introduced auscultation of the chest by means of his invention, the stethoscope.

[15] Gabriel Andral (1797-1876) followed the pathological-anatomical tradition of Morgagni. He is credited with establishing the science of hematology and championed chemical analysis of the blood.

[16] Pierre-Charles-Alexandre Louis (1787-1872), foremost representative of the Paris clinical school, advocated careful clinical observation and analysis by *la méthode numerique*. His study of blood-letting, while not an outright condemnation of the practice, contributed to the abandonment of it.

[17] Auguste François Chomel (1788-1858) was physician to La Charité. It was on his wards that Louis conducted his studies.

[18] Émilie du Châtelet (1706-1749), perhaps best known as a mathematician, translated Newton's *Principia* into French.

Chapter XII

[1] Vitalism holds that life cannot be explained by physical and chemical laws and that some non-material "life force" distinguishes living from non-living things. The nervous and hematological systems were the leading nineteenth-century contenders as the seat of the life force.

[2] Organicism, like vitalism, holds that life cannot be explained by physical and chemical laws. The source of life is thought to be not some vital force, but rather a structure of the body itself.

[3] Humoralism has its roots in the ancient Greek notion that disease is caused by an imbalance in the four humors: phlegm, blood, black bile and yellow bile. This theory was widely accepted until Rudolf Virchow, in the 1850s, brought forth his cellular theory of pathology.

[4] Solidism is the theory that disease is the result of morbid changes in the solid parts of the body; hence, it is opposed to humoralism.

[5] These theories stand in opposition to vitalism; they hold that life is sufficiently explained by chemical and physical laws.

[6] Haller's experimental work had distinguished between irritability and sensibility in the nervous system. Cullen and others used the notion of irritability as a general explanation of disease.

[7] Giovanni Rasori (1762-1837), building on Brown's theory, developed his system of contro-stimulism, or contra-stimulism, which held that since diseases were caused by improper excitement, therapy should aim at neutralizing that excitement by means of contra-stimuli.

[8] See the introduction to this volume, pp. 12-13, for brief descriptions of the doctrines of Cullen, Brown and Broussais.

[9] Homeopathy, founded by Samuel Hahnemann (1755-1843), holds that disease is the result of "psora," or suppressed itch. Therapy consists in administering minute dosages of a drug that in larger amounts would cause the symptoms of the disease.

[10] Hydropathy was founded by Vincenz Preissnitz (1799-1851), who established his first water cure center in 1826 in Silesia. The system involved baths, which were really vehicles for the application of heat or cold, and internal cleansing.

[11] Methodism developed from the teachings of Asclepiades of Prusa in Bithynia, who was born about 124 B.C., and went to Rome as a young man. The methodists attended to the disease alone, and not the individual's medical history and situation.

[12] Thessalus of Tralles (fl. c. A.D. 70-95) was a prominent Roman physician and early advocate of methodism.

[13] Caelius Aurelianus (fl. fifth century A.D.) was a methodist and one of the most prominent physicians of late antiquity. His most famous work was *De morbis acutis et chronicis*.

[14] Jan Baptista Van Helmont (1577-1644) was a Flemish physician. He advocated a complex system of supernatural agencies called *archei*, which are controlled by a

central *archeus* and direct the operations of the body. Diseases are caused by some disturbance of the *archeus*, and so therapy aims at bring the *archeus* back to normal.

[15] These and many other patent medicines proliferated in the nineteenth century; their purveyors made all sorts of extravagant (and false) claims about their efficacy. Patent medicines were satirized by Edgar Allen Poe. See Burton R. Pollin, "Poe's Literary Use of 'Oppodeldoc' and Other Patent Medicines," *Poe Studies* 4 (1971): 30-32.

[16] Lobelia, a poisonous plant native to the eastern United States, is still advocated by herbal therapists as a treatment for asthma and bronchitis. Its discovery is attributed to Samuel Thomson, but it was probably in use in New England long before Thomson.

[17] Original Edition has "*àpriori*."

[18] See the introduction to this volume, p. 13, for a brief description of the doctrine of Benjamin Rush.

[19] John Esten Cooke (1783-1853) published his *Treatise of Pathology and Therapeutics* in 1828.

[20] Joseph Adams Gallup (1769-1849) established the medical college in Woodstock, Vermont. In 1839, he published his *Outlines of the Institutes of Medicine*.

[21] Henry Clutterbuck (1767-1856) published *A Series of Essays on Inflammation and its Varieties; Tending to Show, That Most Diseases Either Consist in Inflammation, or Are Consequences of It, More or Less Remote* in 1846.

[22] *Histoire de Gil Blas de Santillane* is a novel set in Spain, but written in French by Alain-René LeSage (1668-1747).

[23] Thomas Sydenham's *A Treatise of the Gout and Dropsy* was published in 1683.

[24] Sir Henry Holland (1788-1873) was a physician to Queen Victoria as well as a traveler and naturalist.

Chapter XIII

[1] Charles Caldwell (1772-1853) founded the medical department at Transylvania University and moved to the school at Louisville in 1837. A new edition of William Cullen's *First Lines of the Practice of Physic. . . With Notes and Observations, Practical and Explanatory, and a Preliminary Discourse, in the Defence of Classical Medicine, by Charles Caldwell, M.D.* was published in Philadelphia in 1816.

[2] *An Enquiry into the Origin of the Late Epidemic Fever in Philadelphia.*

[3] Edward Miller (1760-1812), along with Samuel L. Mitchill and Elihu Hubbard Smith, began the first American medical journal, *The Medical Repository*, in New York in 1797.

[4] It is possible that Bartlett is simply arguing that Cooke's mechanism fails to explain the vital properties he observes. However, Bartlett is showing vitalist leanings, which would be hard to justify using his own empirical philosophy. For further discussion see Chapter XV.

[5] Uroscopy, the diagnosis of disease by the examination of the urine using sight, smell and taste, was widely practiced throughout Europe during medieval and renaissance times.

[6] Thomas Miner (1777-1841) and William Tully (1785-1859) were Connecticut physicians. They published their *Essays upon Fevers and other Medical Subjects* in 1823.

[7] See the introduction to this volume, p. 13, on Samuel Thomson.

[8] Thomson published many works on his system, but I can find no such title.

[9] Elisha Perkins (1741-1799), a Connecticut physician, invented metallic tractors around 1796. These pairs of instruments were alleged to be of peculiar metallic composition and were used to treat local inflammations, pains of rheumatism, and other diseases.

[10] Original edition has "barreness."

Chapter XIV

[1] Bartlett does not make explicit *why* such structures are primary and essential to classification. He seems to presume that simple observation of some structures makes their importance apparent.

[2] In rejecting the physiological conception of disease, Bartlett adopts an ontological conception of disease—a position that would be vindicated later in the nineteenth century with the acceptance of Virchow's explanation of disease in terms of cellular pathology. The argument that nosology depends upon the metaphysical conception of disease is plausible, but Bartlett's claim that a physiological conception of disease would make any nosological classification impossible is too strong.

[3] Bartlett seems to assume that there can be only one correct classification of disease, whereas nosology, even if based purely on empirical observation, may take different forms depending upon the purpose for which the nosology is constructed. The failure to see this may be a result of Bartlett's failure to make explicit why certain phenomena are "primary and essential" to nosological classification. Later in the chapter, however, he will give examples that demonstrate his principles of nosology.

[4] François Boissier de la Croix de Sauvages (1706-1767) constructed a nosology consisting of over two thousand diseases, with ten classes and forty-two orders. The orders were further divided into genera and species. The classes of disease were based on observable clinical criteria. His *Nosologia methodica sistens morborum classes juxta Sydenhami mentem et botanicorum ordinem* was published in 1768.

[5] William Cullen (1710-1790) published the first edition of his *Synopsis nosologiae methodicae* in 1769. Cullen classified diseases into four categories: fevers, nervous disorders, cachexias, and local diseases.

[6] John Mason Good (1764-1827), an English surgeon, published *A Physiological System of Nosology* in 1823.

[7] Felix Plater, or Platter (1536-1614), was born in Basel, where he practiced medicine and lectured. His observations on many diseases were published in the late sixteenth and early seventeenth centuries.

[8] This idea picks up on Bartlett's point, which he made in the eleventh chapter, about the impossibility of giving necessary and sufficient conditions for defining disease. As he suggested in that chapter, nosology is more akin to grouping according to what would become known as the "family resemblance" of Wittgenstein.

[9] Hives

[10] Cutaneous anthrax

[11] Bartlett here runs headlong into the problem of which phenomena more naturally go together. Since he has nothing but observation to settle this question, he can admit of no answer.

[12] Parkinsonism

[13] That Bartlett should see these disparate diseases as a natural group shows that he takes natural similarities to be nothing more than similarities in observed symptoms and signs.

Chapter XV

[1] Justus von Leibig (1803-1873) is primarily remembered for his work in agricultural chemistry and the role of yeast in fermentation.

[2] Finally, Bartlett shows his skepticism about discovering some ultimate vital principle. This would be in keeping with his overall empiricist project. Nonetheless, he seems sufficiently respectful toward the idea that he does not want to condemn it outright.

[3] Obviously, Bartlett seriously underestimated the potential of the basic sciences to shed light on the phenomena of life.

Chapter XVI

[1] In the nineteenth century, "spotted fever" was used to refer to typhus and to cerebrospinal meningitis; the two diseases were not well differentiated at the time. The rickettsial infection now known as Rocky Mountain Spotted Fever, which is actually more common in the northeast United States, was not described until later in the century.

[2] Exactly the opposite has turned out to be the case. With the discovery of the microorganisms responsible for the acute diseases Bartlett mentions, success has been achieved in treating them. It is now the chronic conditions that prove most problematic in the search for effective therapy and that cause greater morbidity, at least in developed countries. Bartlett expresses an idea worth noting, however. He sees the acute diseases as having etiologies that may forever elude science, and hence forever be uncontrollable. The chronic diseases, on the other hand, are caused by factors that are subject to human control.

[3] Pierre Antoine Prost (d. 1832) was physician and surgeon at the Hôtel Dieu de Lyon. His *Médecine éclairée par l'observation et l'ouverture des corps*, as Bartlett says, was published in 1804.

[4] Jean-Nicholas Corvisart (1755-1821) was Napoleon's personal physician. The first edition of his *Essai sur les maladies et les lésions organiques du coeur et des gros vaisseaux* was published in 1806.

[5] René-Théophile-Hyacinthe Laennec published his *Traité de l'auscultation médiate et des maladies des poumons et du coeur* in 1818.

[6] Jean-Baptiste Bouillaud (1796-1881) published his .*Traité clinique des maladies du coeur: précédé de recherches nouvelles sur l'anatomie et la physiologie de cet organ* in 1835.

[7] Marc Antoine Petit (1762-1840) published *Traité de la fièvre entéro-mésentérique, observée, reconnue et signalée publiquement a l'Hôtel-Dieu de Paris, dans les années 1811, 1812 et 1813*, written with the help of Êtienne Renaud Augustin Serres (1786-1868).

[8] Léon Louis Rostan (1790-1866) published his *Recherches sur une maladie encore peu connue, qui a reçu le nom de ramollissement du cerveau* in 1820.

[9] Jean-André Rochoux (1787-1852) published his *Recherches sur l'apoplexie* in 1814. A second edition, entitled *Recherches sur l'apoplexie, et sur plusieurs autres maladies de l'appareil nerveux cérébro-spinal*, appeared in 1833.

[10] From 1820-1834, Claude François Lallemand (1790-1853) published his *Recherches anatomico-pathologiques sur l'encéphale et ses dépendances*, a series of pathological reports of diseases of the meninges and the brain. He was probably most known, however, for his insistent teaching that loss of sperm was dangerous to health.

[11] Alexandre-Jean-Baptiste Parent-Duchâtelet (1790-1836) published *Recherches sur l'inflammation de l'arachnoïde cérébrale et spinale, ou, Histoire théorique et pratique de l'arachnitis : ouvrage fait conjointement* in 1821.

[12] Louis Martinet (1795-1875) collaborated with Parent-Duchâtelet and also wrote manuals of pathology and therapeutics.

[13] Maxime Durand-Fardel (1815-1899) published *Traité du ramollissement du cerveau* in 1843.

[14] Fernand Martin-Solon (1795-1856) published *De l'albuminurie ou hydropisie causée par maladie des reins; modifications de l'urine dans cet état morbide, a l'epoque critique des maladies aigues et durant le cours de quelques affections bilieuses* in 1838.

[15] Pierre François Olive Rayer (1793-1867) published *Traité des maladies des reins, étudiées en elles-mêmes et dans leurs rapports avec les maladies des uretères, de la vessie, de la prostate, de l'urètre, etc.* in 1837 and his three-volume *Traité des maladies des reins: et des altérations de la sécrétion urinaire, etudiées en elles-mémes et dans leurs rapports avec les maladies de uretères, de la vessie, de la prostate, de l'urèthre, etc.: avec un atlas in-folio* in 1839-1841.

[16] Among the writings of François Louis Isidore Valleix (1807-1855) was his *Traité des névralgies; ou, Affections douloureuses des nerfs*, published in 1841.

[17] Augustin Grisolle (1811-1869) published his *Mémoire sur la pneumonie* in 1836, and his *Traité pratique de la pneumonie : aux différens ages, et dans ses rapports, avec les autres maladies aigues et chroniques* in 1841.

[18] Frédéric Rilliet (1814-1861), with Antoine Charles Ernest de. Barthez (1811-1891), published *Maladies des enfans; affections de poitrine. 1. ptie. Pneumonie* in 1838. *Their Traité clinique et pratique des maladies des enfants* appeared in 1843, with a second French edition in 1853 and a third edition in 1884. German translations, *Handbuch der Kinderkrankheiten*, appeared in 1844 and 1855.

[19] Andral published his *Essai d'hématologie pathologique* in Paris in 1843. In the same year, a longer book by the same title was published by Andral, Gavarret, and Onésime Delafond (1805-1861) in Bruxelles. Gavarret had published his *Recherches sur les modifications de proportion de quelques principes du sang, fibrine, globules, matériaux solides du sérum, et eau, dans les maladies* in 1842.

[20] Shakespeare, *Othello* 1.3.80-81.

[21] George Cleghorn (1716-1794) was a Scottish physician who spent several years in Minorca and wrote about the diseases of that place.

[22] William Heberden (1710-1801) was a keen observer and contributed many papers to the medical literature his *Commentaries on the History and Cure of Diseases* was published in 1802.

[23] Thomas Percival (1740-1804) published several essays and advocated improved public sanitation. His *Medical Ethics* (1803) served as the basis for the first code of ethics of the American Medical Association in 1847.

[24] John Cheyne (1777-1836) was a Scottish physician who published important work on the anatomy and pathology of the larynx.

[25] Thomas Bateman (1778-1821) was one of the founders of the *Edinburgh Medical and Surgical Journal* and wrote especially on diseases of the skin.

[26] Samuel Black (1762-1832) was a native of Northern Ireland. He pioneered the diagnosis and study of angina pectoris.

[27] Bartlett is probably referring to Gulielmus, or William, Woollcombe, whose *Remarks on the Frequency and Fatality of Different Diseases: Particularly on the Progressive Increase of Consumption: With Observations on the Influence of the Seasons on Mortality* was published in 1808.

[28] Given the reference on the following page, Bartlett may be referring to William Brown (1748-1792) and not William Brown, the Younger (1796-1887), Fellow of the Royal College of Surgeons, Edinburgh.

[29] Sir Henry Marsh (1790-1860) was an Irish physician. He became physician to the queen in the 1830s.

[30] Percival's *Philosophical, Medical and Experimental Essays* was published in London in 1776.

[31] This work was published in London in 1836.

[32] Xavier Bichat (1771-1802) published the first edition of his *Anatomie générale : appliquée à la physiologie et à la médecine* in 1801.

[33] See William Osler's "The Influence of Louis on American Medicine."

[34] Charles Caldwell published *An Address to the Philadelphia Medical Society on the Analogies between Yellow Fever and True Plague; Delivered by Appointment, on the 20th of February, 1801.*

[35] Benjamin Rush published his *Medical Inquiries and Observations, Volume 3: An Account of the Bilious Yellow Fever, As It Appeared in the City of Philadelphia, in the Year 1793* in 1794.

[36] Edward Miller published his *Report on the Malignant Disease, Which Prevailed in the City of New-York, in the Autumn of 1805* in 1806. In the same year, a French translation by J. D. Dupont appeared as *Histoire de la maladie maligne appelée fièvre jaune, avec ses effets pendant l'automne de 1805 à New-York, qui prouvent qu'elle n'est nullement contagieuse*.

[37] James Jackson, Jr. (1810-1834) studied with Louis in Paris in 1831-32 before returning to Boston and earning his medical degree at Harvard. He died at the age of twenty-four of dysentery.

[38] William Wood Gerhard (1809-1872) went to Paris in 1831 or 1832 and studied with Chomel, Andral and Louis. He returned to the United States and had a long tenure as professor in the University of Pennsylvania. His *On the Diagnosis of Diseases of the Chest; Based upon the Comparison of Their Physical and General Signs* was published in 1836.

[39] Caspar Wistar Pennock (1799-1867) and W. W. Gerhard are credited with having accurately distinguished between typhus and typhoid fevers during an outbreak in Philadelphia in 1836. See Gerhard's "On the Typhus Fever Which Occurred at Philadelphia in the Spring and Summer of 1836." *American Journal of the Medical*

Sciences 19 (1837):289-322; 20 (1837):289-322. Pennock and Gerhard also co-authored *Observations on the Cholera of Paris*, published in 1832.

[40] Thomas Stewardson (1807-1878), of Philadelphia, was another of the American physicians who went to study in France. He later specialized in treating yellow fever. The black vomiting of bilious remittent fever may have been caused by either malaria or yellow fever.

[41] James Jackson (1777-1867) was a co-founder of the Massachusetts General Hospital and professor at Harvard. Bartlett's notes from Jackson's 1824-1825 lectures are extant (Y2:16:1). In 1838, Jackson published *A Report Founded on the Cases of Typhoid Fever, or the Common Continued Fever of New England, Which Occurred in the Massachusetts General Hospital, from the Opening of That Institution, in September, 1821, to the End of 1835.*

[42] Enoch Hale (1790-1848) published his *.Observations on the Typhoid Fever of New England: Read at the Annual Meeting of the Massachusetts Medical Society, May 29, 1839.*

[43] Thomas Sutton (1767-1835) published his *Tracts on Delirium Tremens, on Peritonitis, and on Some Other Inflammatory Affections, and on the Gout* in London in 1813.

[44] John Ware's (1795-1864) *Remarks on the History and Treatment of Delirium Tremens* was published in Boston in 1831. A revised version was published in 1838 and entitled *On the Treatment of Delirium Tremens; Being an Appendix to an Essay on This Disease, Formerly Published.*

[45] *Secale cornutum*, or ergot, a rye fungus, was used for accelerating parturition going back to the sixteenth century. John Stearns (1770-1848) published his *Observations on the Secale Cornutum, or Ergot : With Directions for Its Use in Parturition* in 1822.

Index

[1] Original edition has "15."

The Philosophy of Therapeutics

[1] In the manuscript, Bartlett has completely crossed out the tenth proposition, and has renumbered the eleventh and twelfth as the tenth and eleventh, respectively. This edition restores the tenth proposition, thus keeping the total number of propositions at twelve.

[2] Here ends the material from manuscripts owned by the University of Rochester Library. What follows is from manuscripts owned by the Yale University Library.

[3] Manuscript has "corrolary."

[4] It seems obvious that such things as chemical structure do have much to do with determining the therapeutic characteristics of any substance. The point that Bartlett seems to be making is that knowledge of the chemical structure, etc. of a substance would not necessarily lead directly to knowledge of its therapeutic efficacy. Even if one reasons hypothetically from a substance's structure to its therapeutic efficacy, no knowledge of therapeutic efficacy would be obtained until the substance was empirically tested and shown to be therapeutically effective.

[5] Manuscript has "emenations."

[6] Bartlett thus seems to accept the existence of the vital force, and even other forces of nature, and to attribute a divine nature to them. As such, they are not susceptible to empirical analysis.

[7] Bartlett asserts an ontological conception of disease. See the fourteenth chapter in the second part of the *Essay*.

[8] Physiological medicine was the system of Broussais. See the introduction to this volume, pp. 12-13, and the twelfth chapter of the second part of Bartlett's *Essay*.

[9] See note 17 in Chapter XVI of the second part of Bartlett's *Essay*.

[10] Manuscript has "corrolary."

[11] Manuscript has "inmitigated."

[12] This paragraph is written in a different hand, but is labeled to be inserted into the manuscript at this point.

[13] Manuscript omits "do."

[14] Manuscript has "untill."

[15] This paragraph and the following two paragraphs appear on separate manuscript fragments, but the sections are numbered to fit into this point in the text.

[16] Nathan Smith (1762-1828) founded the Dartmouth Medical School and also taught at Yale, the University of Vermont and Bowdoin. His son, Nathan Ryno, taught at Transylvania and Maryland and was a colleague of Bartlett.

[17] Peter Mere Latham (1789-1875) and his father John Latham (1761-1843) were both distinguished English physicians.

[18] Manuscript has "feirce."

[19] Manuscript has "loser."

[20] Bartlett is apparently making a metaphysical claim about the complexity of the life sciences. Yet, given Bartlett's previously stated expectation that progress will be made in the life sciences, it is hard to avoid the conclusion that the complexity is the cause of imperfect knowledge. Hence, this is more of an epistemological claim.

[21] This and the next paragraph are crossed out in the manuscript.

[22] Manuscript has "Montaign." Manuscript also omits "know." The source of this quote is unknown. It is certainly in the spirit of Montaigne, but also of Pascal.

[23] This proposition and the following paragraph explaining it are crossed out in the manuscript.

[24] Manuscript has "Eleventh" crossed out and "Tenth" inserted.

[25] Manuscript has "necessary."

[26] See Warner, *The Therapeutic Perspective*, pp. 58-63.

[27] Manuscript has "Twelfth" crossed out and "Eleventh" inserted.

INDEX

Ackerknecht, Erwin, 17, 25
Allan, James S., 20
American medical doctrines, 152-62
anatomy
  definition of, 81-2
  sources of our knowledge, 88-9
Andral, Gabriel, 132, 184, 222, 229, 230
Annan, Samuel, 9, 20, 21
Aristotle, 151
Armstrong, John, 116, 220
atomic theory, 73-4, 211
Auber, Théophile Charles Emmanuel, 52

Bacon, Sir Francis, 58, 63, 84, 146, 150, 211
Baltimore, 8
Barthez, Antoine Charles Ernest de, 184, 229
Bartlett, Caroline, 4
Bartlett, Elisha
  academic appointments, 6-10
  Address Delivered at the Anniversary Celebration of the Birth of Spurzheim and the Organization of the Boston Phrenological Society, January 1, 1838, 33
  biographical sketch, 3-11
  Brief Sketch of the Life, Character, and Writings of Dr. William Charles Wells, delivered before the Louisville Medical Society, December 7, 1849, 34
  Discourse on the Times, Character, and Writings of Hippocrates, 35
  education, 3-5
  Essay on the Philosophy of Medical Science, 2, 8, 9, 16-7, 18-25, 27, 29, 30, 31
  Head and the Heart, or the Relative Importance of Intellectual and Moral Education: A Lecture delivered before the American Institute of Instruction, in Lowell, August, 1838, 36-7
  History, Diagnosis and Treatment of Edematous Laryngitis, 33
  History, Diagnosis, and Treatment of the Fevers of the United States, 32
  History, Diagnosis, and Treatment of Typhoid and of Typhus Fever; With an Essay on the Diagnosis of Bilious Remittent and of Yellow Fever, 32
  Inquiry into the Degree of Certainty of Medicine, and into the Nature and Extent of Its Power over Disease, 26-9
  Introductory Lecture on the Objects and Nature of Medical Science, 17-8
  "Laws of Sobriety," and the "Temperance Reform:" An Address delivered before the Young Men's Temperance Society in Lowell, March 8, 1835, 36
  Lecture on the Sense of the Beautiful, Delivered before the Lexington Lyceum, January 20, 1843, 37
  Medical Magazine, 32
  Monthly Journal of Medical Literature, 32
  Obedience to the Laws of Health, a Moral Duty: A Lecture Delivered before the American Physiological Society, January 30, 1838, 36
  Oration Delivered to the Municipal Authorities and the Citizens of Lowell, July 4, 1848, 37
  Philosophy of Therapeutics, 2, 10, 17, 28, 29-31
  poetry, 37-9
  Sketches of the Character and Writings of Eminent Living Surgeons and Physicians of Paris, 32
  terminal illness, 10-1
  transposition of organs, 10
  Valedictory Address to the Graduating Class of Transylvania University, 1843, 34
  Vindication of the Character and Condition of the Females Employed in the Lowell Mills Against the Charges Contained in the Boston Times and the Boston Quarterly, 37
  William Paley's Natural Theology, 32
Bartlett, George, 3
Bartlett, John, 3
Bartlett, Otis, 3
Bartlett, Waite Buffum, 3
Bateman, Thomas, 185, 229

Bean, William, 25
Beck, J.R., 10
Bell, T.S., 9
Berkshire Medical Institute, 6-7
Bichat, Xavier, 187, 230
Bigelow, Jacob, 4
births, division between sexes, 121-3, 179 n. b
Black, Samuel, 185, 229
Blane, Sir Gilbert, 52, 150 n. e, 180 n. c, 183 n. d, 185, 209
blood-letting, 13, 26, 29, 31, 105, 147
Boott, Francis, 116, 220
Boston, 4, 5, 6, 32, 37, 188, 220, 230, 231
botany, system of classification in, 164-66
Boullaud, Jean-Baptiste, 52, 183, 209, 227
Brewster, Sir David, 68, 211
British medicine, 185-87
Broussais, F. J.V., 12, 145, 145, 167, 183, 223, 232
Brown, John, 12, 31, 145, 147, 223
Brown, William, 185, 186, 230
Brown University, 4

Cabanis, Pierre-Jean-Georges, 14
Caelius (Coelius) Aurelianus, 135, 219
calculation of probabilities, 121-3
Caldwell, Charles, 188, 230
Cassedy, James H., 14, 24, 37
cause and effect, 22, 24, 101, 136, 151, 156, 175, 219
causes, final, 89-90
certainty, 26, 125, 196-99
chemical doctrine, 134, 224
chemistry, its functions, 173-76
Cheyne, John, 132 n. e, 180, 184, 222, 229
Chomel, Aususte François, 132, 180, 184, 222, 230
Clark, Alonzo, 10, 11
classification
  in physical science, 76-78
  of diseases, 163-73
  of phenomena of therapeutics, 200-4
Cleghorn, George, 185, 229
Clutterbuck, Henry, 145, 224
College of Physicians and Surgeons, 7, 10, 29
Collins, Robert, 187
Congestion, a common element, 112
contro-stimulism (contra-stimulism), 134, 223
Cooke, John Esten, 145, 155-57, 224, 225
  his theory of therapeutics, 157-8
Corvisart, Jean Nicholas, 183, 227
country medical colleges, 6
Cragie, David, 186
Cullen, William, 12, 136, 168, 187, 223, 226
  his theory of fevers, 135-36

Cuvier, Georges, 131, 221

Dalton, Dr., his atomic theory, 74
Dartmouth College, 7
Darwin, Erasmus, 116, 221
Davy, Sir Humphrey, 79-80, 212
definitions, 81-83
Delafond, Onésime, 229
delirium tremens, its treatment, 188-9
diagnosis
  reason for its importance, 107-8
  its relation to therapeutics, 107-10, 219
  nosological, 110-16
  therapeutical, 117-8
diamond, its combustibility, 68
disease(s)
  categories, 26-7, 226
  nature of, 113-4, 221, 226, 227, 232
  seat of, 112-3
  systems of explanation, 12-4
  their common elements, 111-2
Dickens, Charles, 1, 10, 37
doctrines, medical, 132-63
  American, 152-63
  general prevalence of, 133-4
  their bad influence, 133-4
Du Châtelet, Émilie, 132, 222, 228
Durand-Fardel, Maxime, 184, 228

Eclectics, 1
Edinburgh, 34, 105, 136
Empirics, 100
empiricism, 16, 21, 25, 30, 36, 152, 211
erroneous notions, extent of, 84, 87, 88
etiology, sources of our knowledge of not deducible from pathology, 98-9

facts, 18, 19, 22, 23, 24, 62-8, 83-87, 199, 209, 210, 212, 217, 218
fever
  Dr. Rush's theory of, 152-3
  inflammatory, a common element, 111, 112
  typhoid and typhus, their differences, 115-18
  typhoidal, a common element, 111, 112

forces, chemical and vital, 173-77
Frauenhofer, Joseph von, 75, 212
French medical observation, 183-5
Fresnel, Augustin Jean, 66-8, 210

Gallup, Joseph Adams, his theory, 159-60, 224
Gavarret, Louis-Dominique-Jules, 121, 122, 124, 126, 132 n. e, 184, 221, 229
Gerhard, William Wood, 188-9, 230

germination, its phenomena, 85-86
Gilman, C.R., 9
Good, John Mason, 168-9, 226
Gould, Augustus Addison, 9
gravitation, its nature or essence, 76-7
Green, John, 4
Green, John Orne, 6, 7, 8
Gregory, James Craufurd, 186
Grisolle, Augustin, 184, 198, 203, 229

Hahnemann, Samuel, 136-42, 223
Hale, Enoch, 188, 231
Haller, Albrecht von, 132, 216, 222
Hanover, N.H., 7
Heberden, William, 185, 229
Henderson, William, 186
Herschel, Sir John, 66, 210
Heywood, B.F., 4
Hillary, William, 52
Hippocrates, 35, 132, 222
Holland, Henry, 151, 186 n. n, 224
Holmes, Oliver Wendell, 1, 7, 11, 15, 19, 20, 38, 142-4
homeopathy, 1, 13, 27, 134, 136-42, 223
humoralists, 134, 223
Hunter, John, 132, 163, 222
Huxham, John, 185, 221
Huygens, Christiaan, 75, 212
hydropathy, 134, 223
hypothesis, 18, 19, 22, 23, 24, 71-80, 134
  how far admissible in medical science, 150-1
  in physical science, 72-80, 211

inflammation
  sources of our knowledge of, 95-97
  its therapeutical relations, 101-4
irritability doctrine, 134
Irving, Washington, 37

Jackson, James, 4, 19-20, 188, 231
Jackson, Jr., James, 188, 230
Jefferson, Thomas, 16
Jussieu, his classification, 164-66

Kimball, Gilman, 9
King, Lester, 1, 16, 24-5, 29
knowledge, its incompleteness, 72, 108, 167, 196

Laennec, René-Théophile-Hyacinthe, 14, 132, 183, 188, 222, 227
Lagrange, Joseph-Louis, 67, 210
Lallemand, Claude François, 184, 228
Lanza, Vincenzo, 52, 209
Laplace, Pierre, 67, 210
Latham, John, 203, 233
Latham, Peter Mere, 203, 233
law, 18, 22, 24, 39, 69-71, 177, 211, 218
  absolute vs. ascertainable, 118-20, 220, 221
  of great numbers, 121-2
  of therapeutics, 205
Lawson, L.M, 22, 27
Lee, Charles A., 21
Leghorn, Italy, 5
Leibig, Justus von, 173-5, 227
Leigh, Edwin, 23
Lexington, Ky., 8, 9
Linnaeus, his classification, 164-67
London, 5, 16, 187, 189
Louis, Pierre-Charles-Alexandre, 9, 14, 15, 132, 132 n. e, 183, n. d, 184-5, 186, 187, 188, 209, 222, 230
Louisville, Ky., 9-10
Lowell, Mass., 1, 5-6

Malus, Etienne-Louis, 66, 210
Mann, Horace, 7
marble, its properties and relations, 63-5
Marsh, Sir Henry, 185, 230
Martinet, Louis, 184, 228
Martin-Solon, Fernand, 184, 228
mathematics, functions of, 66-7
matter, constitution of, 72-3, 80-2
mechanical doctrine, 134
medical science
  its future prospects, 176-77
  object of, 17-8
  progress, 178
  propositions in, 84
medicine, patent, 224
method, numerical, 132 n. e, 184-5, 186
methodism, 134-36, 223
Miller, Edward, 188, 230
Miner, Thomas, his theory, 159-60, 225
Morgagni, Giovanni Battista, 132, 222
Mossotti, Giovanni Fabrizio, 212

Newton, Sir Isaac, 66, 67, 68, 74, 75, 76, 78, 79, 210
Nine Partners, N.Y., 3
nosologies
  systematic, 163-73
  their defects, 167-9
Nott, Josiah C., 23

object of essay, 60
observation
  its difficulties, 144
  its imperfections, 127-8
  the sole foundation of physical science, 62-3

the sole foundation of medical science, 83-107
optics
sources of our knowledge of, 66-8
theories of, 74-6
organicism, 134, 222
Osler, William, 1, 15, 17, 230

Palmer, B.R., 11
Parent-Duchâtelet, Alexandre-Jean-Baptiste, 184, 228
Paris, 4, 9, 14, 15, 16, 66, 121, 122, 187, 189
Paris clinical school, 1, 9, 12, 19, 25
in America, 12-4
Pasteur, Louis, 14, 28
pathology
definition of, 87
its laws, 123-4
not deducible from physiology, 92-98, 217
qualifications, 92-5
relations of, 83
Pennock, Caspar Wistar, 188, 230
Percival, Thomas, 185, 186, 229, 230
Perkins, Elisha, 225
Peter, Robert, 9
Petit, Marc Antoine, 183, 228
phenomena, invariableness of, 69-70
philosophy, false, prevalence of, 83-5
philosophy of medicine
nineteenth century, 11-6
in America, 12-4
phrenology, 33-4
physical science
independent of hypothesis, 72-3
its laws, or principles, 69-71
nature and elements of, 61-2
propositions in, 59
physiology
definition of, 82
not deducible from anatomy, 89-92
Pinel, Philippe, 14
Pittsfield, Mass., 6-7
Plater (Platter), Felix, 226
Plato, 151
Poe, Edgar Allen, 224
principles in medical science, 118-32
contingent nature of their elements, 130-1
how ascertained, 119-27
Pringle, Sir John, 104 n. c, 116, 185, 221
Prost, Pierre Antoine, 183, 227
Providence, R.I., 4

Quakers, 3

Rasori, Giovanni, 136, 223
rationalists, 100
Rayer, Pierre François Olive, 184, 228
reasoning
a priori, its functions, 68
deductive, 147-8, 217
inductive, 210, 211
Reid, John, 186
relationships, invariable, 69-70
remedies
action of on lower animals, 106-7
new, 178-9
respiration, its phenomena, 86-7
Rilliet, Frédéric, 184, 229
Roby, 11, 20
Rochoux, Jean-André, 183, 228
Rostan, Léon Louis, 183, 228
Rousseau, Jean-Jacques, 96, 182, 218
Rush, Benjamin, 13, 25, 167 n. a, 188, 224
his theories, 152-4

Sauvages, François Bossier de la Croix de, 168, 226
school of observation
American, 187-89
British, 185-87
French, 183-85
Serres, Êtienne Renaud Augustin, 183, 203
sexes, proportion of 121-23
Shearman, J.H., 28
Silliman, Benjamin, 7, 10, 11
Slater, Elizabeth, 5
Slater, John, 5
Slater, Samuel, 5
Smith, Nathan, 203, 232
Smithfield, R.I., 3, 10, 11
solidism, 134, 219
statistics, medical, 121-27
Stearns, John, 189, 231
Stewardson, Thomas, 231
Stewart, Alexander, 186
Stillé, Albert, 27
Sutton, Thomas, 188, 231
Sydenham, Thomas, 185, 187, 216, 224
his philosophy, 147-8

theory, 18, 19, 23, 24, 210
therapeutics, 2, 19, 23, 29-31
character of, 204-5
definition of, 193-4
indication for, 206
its laws, or principles, 124-27
modus operandi, 206-7
not deducible from pathology, 100-4, 199-200, 212-3
not deducible from physiology, 105-6, 200
object of, 207-8

philosophy of, 193-208
Thessalus, 135, 223
Thomson, Samuel, 13, 224, 225
his theory, 160-2
Thomsonians, 1, 13
Transylvania University, 8, 9, 17
Tully, William, his theory, 159-60, 225

Unitarian Church 3
United States, medicine in, 187-9
University of Maryland, 8
University of the City of New York, 7, 10
Uxbridge, Mass., 3, 4

Valleix, François Louis Isadore, 184, 228
vital forces, 18, 173, 227, 232
vitalism, 134, 222, 223

Ware, John, 188, 227
Warner, John Harley, 24, 233
Wheaton, Levi, 4
Whewell, William, 73, 172, 174, 175, 211
Willard, George, 4
Wollaston, William, 68, 210
Woodstock, Vt., 9
Woolcombe, William, 132 n. e, 185, 186, 229
Worcester, Mass., 4

Young, Thomas, 67, 210

Philosophy and Medicine

1. H. Tristram Engelhardt, Jr. and S.F. Spicker (eds.): *Evaluation and Explanation in the Biomedical Sciences.* 1975 ISBN 90-277-0553-4
2. S.F. Spicker and H. Tristram Engelhardt, Jr. (eds.): *Philosophical Dimensions of the Neuro-Medical Sciences.* 1976 ISBN 90-277-0672-7
3. S.F. Spicker and H. Tristram Engelhardt, Jr. (eds.): *Philosophical Medical Ethics.* Its Nature and Significance. 1977 ISBN 90-277-0772-3
4. H. Tristram Engelhardt, Jr. and S.F. Spicker (eds.): *Mental Health.* Philosophical Perspectives. 1978 ISBN 90-277-0828-2
5. B.A. Brody and H. Tristram Engelhardt, Jr. (eds.): *Mental Illness.* Law and Public Policy. 1980 ISBN 90-277-1057-0
6. H. Tristram Engelhardt, Jr., S.F. Spicker and B. Towers (eds.): *Clinical Judgment.* A Critical Appraisal. 1979 ISBN 90-277-0952-1
7. S.F. Spicker (ed.): *Organism, Medicine, and Metaphysics.* Essays in Honor of Hans Jonas on His 75th Birthday. 1978 ISBN 90-277-0823-1
8. E.E. Shelp (ed.): *Justice and Health Care.* 1981 ISBN 90-277-1207-7; Pb 90-277-1251-4
9. S.F. Spicker, J.M. Healey, Jr. and H. Tristram Engelhardt, Jr. (eds.): *The Law-Medicine Relation.* A Philosophical Exploration. 1981 ISBN 90-277-1217-4
10. W.B. Bondeson, H. Tristram Engelhardt, Jr., S.F. Spicker and J.M. White, Jr. (eds.): *New Knowledge in the Biomedical Sciences.* Some Moral Implications of Its Acquisition, Possession, and Use. 1982 ISBN 90-277-1319-7
11. E.E. Shelp (ed.): *Beneficence and Health Care.* 1982 ISBN 90-277-1377-4
12. G.J. Agich (ed.): *Responsibility in Health Care.* 1982 ISBN 90-277-1417-7
13. W.B. Bondeson, H. Tristram Engelhardt, Jr., S.F. Spicker and D.H. Winship: *Abortion and the Status of the Fetus.* 2nd printing, 1984 ISBN 90-277-1493-2
14. E.E. Shelp (ed.): *The Clinical Encounter.* The Moral Fabric of the Patient-Physician Relationship. 1983 ISBN 90-277-1593-9
15. L. Kopelman and J.C. Moskop (eds.): *Ethics and Mental Retardation.* 1984 ISBN 90-277-1630-7
16. L. Nordenfelt and B.I.B. Lindahl (eds.): *Health, Disease, and Causal Explanations in Medicine.* 1984 ISBN 90-277-1660-9
17. E.E. Shelp (ed.): *Virtue and Medicine.* Explorations in the Character of Medicine. 1985 ISBN 90-277-1808-3
18. P. Carrick: *Medical Ethics in Antiquity.* Philosophical Perspectives on Abortion and Euthanasia. 1985 ISBN 90-277-1825-3; Pb 90-277-1915-2
19. J.C. Moskop and L. Kopelman (eds.): *Ethics and Critical Care Medicine.* 1985 ISBN 90-277-1820-2
20. E.E. Shelp (ed.): *Theology and Bioethics.* Exploring the Foundations and Frontiers. 1985 ISBN 90-277-1857-1

Philosophy and Medicine

21. G.J. Agich and C.E. Begley (eds.): *The Price of Health.* 1986 ISBN 90-277-2285-4
22. E.E. Shelp (ed.): *Sexuality and Medicine.* Vol. I: Conceptual Roots. 1987 ISBN 90-277-2290-0; Pb 90-277-2386-9
23. E.E. Shelp (ed.): *Sexuality and Medicine.* Vol. II: Ethical Viewpoints in Transition. 1987 ISBN 1-55608-013-1; Pb 1-55608-016-6
24. R.C. McMillan, H. Tristram Engelhardt, Jr., and S.F. Spicker (eds.): *Euthanasia and the Newborn.* Conflicts Regarding Saving Lives. 1987 ISBN 90-277-2299-4; Pb 1-55608-039-5
25. S.F. Spicker, S.R. Ingman and I.R. Lawson (eds.): *Ethical Dimensions of Geriatric Care.* Value Conflicts for the 21th Century. 1987 ISBN 1-55608-027-1
26. L. Nordenfelt: *On the Nature of Health.* An Action-Theoretic Approach. 2nd, rev. ed. 1995 ISBN 0-7923-3369-1; Pb 0-7923-3470-1
27. S.F. Spicker, W.B. Bondeson and H. Tristram Engelhardt, Jr. (eds.): *The Contraceptive Ethos.* Reproductive Rights and Responsibilities. 1987 ISBN 1-55608-035-2
28. S.F. Spicker, I. Alon, A. de Vries and H. Tristram Engelhardt, Jr. (eds.): *The Use of Human Beings in Research.* With Special Reference to Clinical Trials. 1988 ISBN 1-55608-043-3
29. N.M.P. King, L.R. Churchill and A.W. Cross (eds.): *The Physician as Captain of the Ship.* A Critical Reappraisal. 1988 ISBN 1-55608-044-1
30. H.-M. Sass and R.U. Massey (eds.): *Health Care Systems.* Moral Conflicts in European and American Public Policy. 1988 ISBN 1-55608-045-X
31. R.M. Zaner (ed.): *Death: Beyond Whole-Brain Criteria.* 1988 ISBN 1-55608-053-0
32. B.A. Brody (ed.): *Moral Theory and Moral Judgments in Medical Ethics.* 1988 ISBN 1-55608-060-3
33. L.M. Kopelman and J.C. Moskop (eds.): *Children and Health Care.* Moral and Social Issues. 1989 ISBN 1-55608-078-6
34. E.D. Pellegrino, J.P. Langan and J. Collins Harvey (eds.): *Catholic Perspectives on Medical Morals.* Foundational Issues. 1989 ISBN 1-55608-083-2
35. B.A. Brody (ed.): *Suicide and Euthanasia.* Historical and Contemporary Themes. 1989 ISBN 0-7923-0106-4
36. H.A.M.J. ten Have, G.K. Kimsma and S.F. Spicker (eds.): *The Growth of Medical Knowledge.* 1990 ISBN 0-7923-0736-4
37. I. Löwy (ed.): *The Polish School of Philosophy of Medicine.* From Tytus Chałubiński (1820–1889) to Ludwik Fleck (1896–1961). 1990 ISBN 0-7923-0958-8
38. T.J. Bole III and W.B. Bondeson: *Rights to Health Care.* 1991 ISBN 0-7923-1137-X

Philosophy and Medicine

---

39. M.A.G. Cutter and E.E. Shelp (eds.): *Competency*. A Study of Informal Competency Determinations in Primary Care. 1991 ISBN 0-7923-1304-6
40. J.L. Peset and D. Gracia (eds.): *The Ethics of Diagnosis*. 1992 ISBN 0-7923-1544-8
41. K.W. Wildes, S.J., F. Abel, S.J. and J.C. Harvey (eds.): *Birth, Suffering, and Death*. Catholic Perspectives at the Edges of Life. 1992 [CSiB-1] ISBN 0-7923-1547-2; Pb 0-7923-2545-1
42. S.K. Toombs: *The Meaning of Illness*. A Phenomenological Account of the Different Perspectives of Physician and Patient. 1992 ISBN 0-7923-1570-7; Pb 0-7923-2443-9
43. D. Leder (ed.): *The Body in Medical Thought and Practice*. 1992 ISBN 0-7923-1657-6
44. C. Delkeskamp-Hayes and M.A.G. Cutter (eds.): *Science, Technology, and the Art of Medicine*. European-American Dialogues. 1993 ISBN 0-7923-1869-2
45. R. Baker, D. Porter and R. Porter (eds.): *The Codification of Medical Morality*. Historical and Philosophical Studies of the Formalization of Western Medical Morality in the 18th and 19th Centuries, Volume One: Medical Ethics and Etiquette in the 18th Century. 1993 ISBN 0-7923-1921-4
46. K. Bayertz (ed.): *The Concept of Moral Consensus*. The Case of Technological Interventions in Human Reproduction. 1994 ISBN 0-7923-2615-6
47. L. Nordenfelt (ed.): *Concepts and Measurement of Quality of Life in Health Care*. 1994 [ESiP-1] ISBN 0-7923-2824-8
48. R. Baker and M.A. Strosberg (eds.) with the assistance of J. Bynum: *Legislating Medical Ethics*. A Study of the New York State Do-Not-Resuscitate Law. 1995 ISBN 0-7923-2995-3
49. R. Baker (ed.): *The Codification of Medical Morality*. Historical and Philosophical Studies of the Formalization of Western Morality in the 18th and 19th Centuries, Volume Two: Anglo-American Medical Ethics and Medical Jurisprudence in the 19th Century. 1995 ISBN 0-7923-3528-7; Pb 0-7923-3529-5
50. R.A. Carson and C.R. Burns (eds.): *Philosophy of Medicine and Bioethics*. A Twenty-Year Retrospective and Critical Appraisal. 1997 ISBN 0-7923-3545-7
51. K.W. Wildes, S.J. (ed.): *Critical Choices and Critical Care*. Catholic Perspectives on Allocating Resources in Intensive Care Medicine. 1995 [CSiB-2] ISBN 0-7923-3382-9
52. K. Bayertz (ed.): *Sanctity of Life and Human Dignity*. 1996 ISBN 0-7923-3739-5
53. Kevin Wm. Wildes, S.J. (ed.): *Infertility: A Crossroad of Faith, Medicine, and Technology*. 1996 ISBN 0-7923-4061-2
54. Kazumasa Hoshino (ed.): *Japanese and Western Bioethics*. Studies in Moral Diversity. 1996 ISBN 0-7923-4112-0

Philosophy and Medicine

55. E. Agius and S. Busuttil (eds.): *Germ-Line Intervention and our Responsibilities to Future Generations.* 1998 ISBN 0-7923-4828-1
56. L.B. McCullough: *John Gregory and the Invention of Professional Medical Ethics and the Professional Medical Ethics and the Profession of Medicine.* 1998 ISBN 0-7923-4917-2
57. L.B. McCullough: *John Gregory's Writing on Medical Ethics and Philosophy of Medicine.* 1998 [CoME-1] ISBN 0-7923-5000-6
58. H.A.M.J. ten Have and H.-M. Sass (eds.): *Consensus Formation in Healthcare Ethics.* 1998 [ESiP-2] ISBN 0-7923-4944-X
59. H.A.M.J. ten Have and J.V.M. Welie (eds.): *Ownership of the Human Body.* Philosophical Considerations on the Use of the Human Body and its Parts in Healthcare. 1998 [ESiP-3] ISBN 0-7923-5150-9
60. M.J. Cherry (ed.): *Persons and Their Bodies.* Rights, Responsibilities, Relationships. 1999 ISBN 0-7923-5701-9
61. R. Fan (ed.): *Confucian Bioethics.* 1999 [ASiB-1] ISBN 0-7923-5723-X
62. L.M. Kopelman (ed.): *Building Bioethics.* Conversations with Clouser and Friends on Medical Ethics. 1999 ISBN 0-7923-5853-8
63. W.E. Stempsey: *Disease and Diagnosis.* 2000 PB ISBN 0-7923-6322-1
64. H.T. Engelhardt (ed.): *The Philosophy of Medicine.* Framing the Field. 2000 ISBN 0-7923-6223-3
65. S. Wear, J.J. Bono, G. Logue and A. McEvoy (eds.): *Ethical Issues in Health Care on the Frontiers of the Twenty-First Century.* 2000 ISBN 0-7923-6277-2
66. M. Potts, P.A. Byrne and R.G. Nilges (eds.): *Beyond Brain Death.* The Case Against Brain Based Criteria for Human Death. 2000 ISBN 0-7923-6578-X
67. L.M. Kopelman and K.A. De Ville (eds.): *Physician-Assisted Suicide.* What are the Issues? 2001 ISBN 0-7923-7142-9
68. S.K. Toombs (ed.): *Handbook of Phenomenology and Medicine.* 2001 ISBN 1-4020-0151-7; Pb 1-4020-0200-9
69. R. ter Meulen, W. Arts and R. Muffels (eds.): *Solidarity in Health and Social Care in Europe.* 2001 ISBN 1-4020-0164-9
70. A. Nordgren: *Responsible Genetics.* The Moral Responsibility of Geneticists for the Consequences of Human Genetics Research. 2001 ISBN 1-4020-0201-7
71. J. Tao Lai Po-wah (ed.): *Cross-Cultural Perspectives on the (Im)Possibility of Global Bioethics.* 2002 [ASiB-2] ISBN 1-4020-0498-2
72. P. Taboada, K. Fedoryka Cuddeback and P. Donohue-White (eds.): *Person, Society and Value.* Towards a Personalist Concept of Health. 2002 ISBN 1-4020-0503-2
73. J. Li: *Can Death Be a Harm to the Person Who Dies?* 2002 ISBN 1-4020-0505-9

Philosophy and Medicine

74. H.T. Engelhardt, Jr. and L.M. Rasmussen (eds.): *Bioethics and Moral Content: National Traditions of Health Care Morality*. Papers dedicated in tribute to Kazumasa Hoshino. 2002 ISBN 1-4020-6828-2
75. L.S. Parker and R.A. Ankeny (eds.): *Mutating Concepts, Evolving Disciplines: Genetics, Medicine, and Society*. 2002 ISBN 1-4020-1040-0
76. W.B. Bondeson and J.W. Jones (eds.): *The Ethics of Managed Care: Professional Integrity and Patient Rights*. 2002 ISBN 1-4020-1045-1
77. K.L. Vaux, S. Vaux and M. Sternberg (eds.): *Covenants of Life. Contemporary Medical Ethics in Light of the Thought of Paul Ramsey*. 2002 ISBN 1-4020-1053-2
78. G. Khushf (ed.): *Handbook of Bioethics: Taking Stock of the Field from a Philosophical Perspective*. 2003 ISBN 1-4020-1870-3; Pb 1-4020-1893-2
79. A. Smith Iltis (ed.): *Institutional Integrity in Health Care*. 2003 ISBN 1-4020-1782-0
80. R.Z. Qiu (ed.): *Bioethics: Asian Perspectives A Quest for Moral Diversity.* 2003 [ASiB-3] ISBN 1-4020-1795-2
81. M.A.G. Cutter: *Reframing Disease Contextually*. 2003 ISBN 1-4020-1796-0
82. J. Seifert: *The Philosophical Diseases of Medicine and Their Cure*. Philosophy and Ethics of Medicine, Vol. 1: Foundations. 2004 ISBN 1-4020-2870-9
83. W.E. Stempsey (ed.): *Elisha Bartlett's Philosophy of Medicine*. 2005 [CoME-2] ISBN 1-4020-3041-X
84. C. Tollefsen (ed.): *John Paul II's Contribution to Catholic Bioethics*. 2005 [CSiB-3] ISBN 1-4020-3129-7
85. C. Kaczor: *The Edge of Life*. Human Dignity and Contemporary Bioethics. 2005 [CSiB-4] ISBN 1-4020-3155-6

KLUWER ACADEMIC PUBLISHERS – DORDRECHT / BOSTON / LONDON